The Counter-Counterinsurgency Manual

The Counter-Counterinsurgency Manual:
Or, Notes on Demilitarizing American Society

Network of Concerned Anthropologists
Steering Committee

Catherine Besteman
Andrew Bickford
Greg Feldman
Roberto González
Hugh Gusterson
Gustaaf Houtmann

Jean Jackson
Kanhong Lin
Catherine Lutz
David Price
David Vine

PRICKLY PARADIGM PRESS
CHICAGO

© 2009 Network of Concerned Anthropologists
All rights reserved.

Prickly Paradigm Press, LLC
5629 South University Avenue
Chicago, Il 60637

www.prickly-paradigm.com

ISBN-13: 978-0-9794057-4-7
LCCN: 2009923954

Printed in the United States of America on acid-free paper.

Royalties from the sale of this book will be donated to the Network of Concerned Anthropologists and Iraq Veterans Against the War.

Table of Contents

Preface
  Marshall Sahlins i

Introduction: War, Culture, and Counterinsurgency
  Roberto González, Hugh Gusterson, David Price 1

Part I: Counter-histories of Militarism

  1. The Military Normal
     Catherine Lutz 23

  2. Militarizing Knowledge
     Hugh Gusterson 39

Part II: Countering the Counterinsurgency Manual

  3. Faking Scholarship
     David Price 59

  4. Radical or Reactionary?
     Greg Feldman 77

Part III: Countering Counterinsurgency

  5. Embedded
     Roberto González 97

  6. Counter AFRICOM
     Catherine Besteman 115

Part IV: Anthropological Implications

  7. Anthropology and HUMINT
     Andrew Bickford 135

  8. About Face!
     Kanhong Lin 153

Part V: Alternatives

  9. Proposals for a Humanpolitik
     David Vine 173

Preface

Marshall Sahlins

The plagiarism is symptomatic of the claims of the so-called "military intellectuals" to rewrite the book on counterinsurgency. In the same way that the authors of the US Army and Marine Corps' *Counterinsurgency Field Manual* stake their academic credentials on a pastiche of platitudes taken without acknowledgment from social science sources, their own text is devious and dubious throughout. Complicit in this lack of intellectual integrity, the University of Chicago Press, which published a trade edition of the *Manual* in 2007, comes in for its fair share of disrepute; but no matter, sales large enough to make some Best Sellers' lists afforded the Press the sufficient consolation of profit without honor in its own country.

Flogging the *Counterinsurgency Field Manual* to the public is a telltale clue to the source of its greatest duplicity: that it is also part of a clandestine battle in the *Heimatland*. Ever since the fateful debates among the citizens of Athens during the Peloponnesian War about how to deal with rebellious allies (*cum* imperial subjects), the counterinsurgency wars of democratic republics have been fought on two fronts. Besides the insurgents themselves, the US military is engaged with a potentially formidable opponent of which the *Manual* seems well aware but never mentions: the general will of the American people. Of course, it is sufficiently repeated elsewhere in military writing that the Vietnam War was inadequately fought and ultimately lost because of mounting opposition at home: the decisive "hearts and minds" the US strategy failed to win. In connection with the ongoing Iraq War, the threat of domestic political opposition has been aggravated by the ever-shifting and disingenuous reasons the erstwhile Commander-in-Chief (George W. Bush) put out for why so many Americans and Iraqis must die, including the recurrent conflation of counterinsurgency with a "global war on terror." So now comes the public offering of a field manual that de-emphasizes the combat and bypasses the politics for an anodyne depiction of counterinsurgency as a global project of applied anthropology.

But as demonstrated in the present *Counter-Counterinsurgency Manual*, the truly good news is that the military's appropriation of anthropological theory is incoherent, simplistic and outmoded—not to mention tedious—even as its ethnographic protocols for learning the local cultures amount to an unworkable fantasy. According to the *Manual*, the military staffs and in some measure personnel at all echelons are supposed to gather data in their area of operations on the local peoples' social structures, cultures, institutions and organizations, groups and the relationships between them, languages, myths, systems of authority and power, values, social networks, historical charters, beliefs ("core," "intermediate" and "peripheral"), symbols, ceremonies, ideologies, categories of thought, codes of behavior, norms, taboos, etc., etc. Fact is, it would take many anthropologists with years of training and even more years of fieldwork to do this, by which time pretty much everything would be different, especially at the tactically relevant level of appearances. (Among the other things the military intelligentsia doesn't understand about anthropology is that it is the only discipline apart from high energy physics that studies disappearing objects.) Another fact is that all the military really wants is some superficial handles on the local culture that will allow them to manipulate the people for better or for worse. As deconstructed in the *Counter-Counterinsurgency Manual*, the applied anthropology of the US military may be described something as follows: a planetary strategy of research and destroy, involving the deployment of armed and largely culturally-illiterate American forces from among the thousand or so garrisons now distributed on foreign soil, sometimes complemented by second rate mercenary academics, all charged with an investigation of the cultures of the local peoples sufficient to determine if and how they can be subjugated or, failing that, taken out.

Here is anthropology as a weapon in dubious battles, as the critics rightly claim. For as it is put by a certain Lt. Colonel cited in the counterinsurgency manual:

> There will be no peace.... The de facto role of the US armed forces will be to keep the world safe for our economy and open to our cultural assault. To these ends, we will do a fair amount of killing. We are building an information-based military to do that killing.

But then, whose side are you on, Petraeus? Although the counterinsurgency manual pretends to be based on up-to-date social science, it lacks the critical reflexivity of the latter, since what it dare not address is the Americans' own presence as an invading and occupying power. This silence is intelligible given the political implications of the *Manual*'s historical examples of insurgency—almost all ideologically reactionary—and its covering determination thereof as a sustained uprising against an established government—taken a priori to be a bad thing. There are no pro-insurgency doctrines here: no reflections on American solidarity with the French Resistance during the Nazi occupation; or the CIA's support of the Muslim forces (including a certain Osama bin Laden) fighting the Soviet occupation of Afghanistan. The apparent exception that proves the rule: a programmatic Appendix entitled "A Guide for Action," which closely parallels without acknowledgement a text by T.E. Lawrence on how to promote Arab uprising against Ottoman rule—but is here perversely and deviously turned into a primer of countersurgerncy. One gains the distinct impression that were the *Counterinsurgency Field Manual* available, copies of it could have been found in the haversacks of the British troops of Lord Cornwall engaged in suppressing the American Revolution.

Still, a worse anachronism is that the counterinsurgency manual in the modern American soldier's kit is indeed derived from British colonial practice. Counterinsurgency doctrine, writes Sarah Sewall in a sympathetic Introduction to the Chicago edition of the *Manual*, "is based on principles learned during Britain's early period of imperial policing and relearned during responses to twentieth century independence struggles in Malaya and Kenya." The *Counterinsurgency Field Manual* also draws salutary lessons from Napoleon's occupation of Spain and his failure to cope with the liberation struggle of the Spanish *guerrillas*, as well as from French strategies of colonial warfare in Algeria, Madagascar and Indochina. But we too have ancestors. The *Counterinsurgency Field Manual* follows on a long history of American intervention and suppression of popular uprising in Cuba, the Philippines, Vietnam, El Salvador, Nicaragua, the Dominican Republic and the territory now known as the United States. The only novelty is the explicit "cultural assault": the military intellectuals' quixotic programs for culturally leveraging a Pax Americana by embedding uniformed anthropologists in combat and intelligence operations of an imperialist vintage. If an unwanted effect has been the spawning of appropriately novel networks of resistance, such as the Network of Concerned Anthropologists represented by the authors of this pamphlet, it is because for them the integrity of other peoples' existence is at once an intellectual premise of their discipline and its ethical imperative. They will not put the peoples with whom they live and work at risk of bodily harm, foreign domination or cultural disintegration.

Yet inasmuch as the American national interest is the decisive strategic interest of counterinsurgency practice, morality in respect of the local population is always subject to expediency. The only apparent exception is when tactical advantage can be gained from the humanitarian treatment of certain groups, in which case morality itself is expedient, if still heedless of justice. The principal role of academics in the service of counterinsurgency is to develop the human intelligence (HUMINT) that will allow a triage between those elements of the population to be attacked (or assassinated) and those it would be better not to—in brief, sophisticated targeting.* But in these matters, stability is privileged over justness or goodness, as Sarah Sewall observes in her brilliantly tortured defense of the *Counterinsurgency Field Manual*. She continues:

> It is painful, then, to see the field manual grapple with the lessons from Iraq because the manual can't state the obvious: that imposing a revolution from outside provides a weak and illegitimate basis from which to defeat an insurgency…. Once committed, the United States is harnessed to a beast that may prove impossible to tame. U.S. actions aim to enhance the host government. But what if the government isn't good or brave or wise?

And if the government isn't good, then its enemies, who are by that token US enemies, could very

---

* The curiously pecuniary terms of military newspeak, counterinsurgency and HUMINT, are the only (unwitting) suggestions in the counterinsurgency manual of the one tactic that seems to have had modest success in Iraq: the bribery of erstwhile enemies and uncertain friends. For its part, counterinsurgency bribery felicitously combines the two great driving forces and discourses of the Pax Americana: militarism and neoliberalism. But as demonstrated in Iraq, the good news here is that the costs of each, militarism and neoliberalism, are much greater than the benefits from the other.

well be. The problem becomes the more painful because a major effect of the American interventions in Iraq—as can be expected anywhere a coup d'etat by a foreign force is then "legitimated" by the installation of a proxy "democratic" regime—has been to turn a plural society of different ethnic, tribal and religious groups into a factionalized state of nature: that is, by remaking their differences into unconditional causes-to die-for in the ensuing all-out competition for political and economic power. "Stuff happens." But the enlistment of human scientists in counterinsurgency saves lives, the apologists of the practice have argued, and anything that saves lives is a good thing. Yet aside from the fact that better targeting merely implies that more of certain kinds of people are killed and fewer of other kinds, under the circumstances where American interests are privileged over local interests, where expedience trumps justice and a myriad of factions are joined in uncompromising battle for wealth and power, nothing guarantees that the people who are killed in counterinsurgency operations are the right kinds.

These observations and many more are made in detail in the pages of that follow. But above all the authors and the Network of Concerned Anthropologists deserve our gratitude and admiration for defending the better values of their discipline and their nation by giving those who had forgotten it to understand that other peoples were not born to live and die for our sake.

Marshall Sahlins
Chicago, December 2008

Introduction
War, Culture, and Counterinsurgency
Roberto González, Hugh Gusterson, and David Price

Dispatches

1. From the *Village Voice*, March 4, 2003:

> NEW YORK, New York—While Pentagon war planners may be gunning for an attack on Iraq by mid March, heavily armed soldiers have already quietly seized a strategic position: your Easter basket. National retailers like Kmart and Walgreens have stocked their shelves with baskets in which the traditional chocolate rabbit centerpiece has been displaced by plastic military action figures and their make-believe lethal paraphernalia...

At the Astor Place Kmart, the encampment is on display just inside the main entrance. A camouflaged sandy-haired soldier with an American-flag arm patch stands alert in a teal, pink, and yellow basket beneath a pretty green-and-purple bow. Within a doll-arm's reach are a machine gun, rifle, hand grenade, large knife, pistol, and round of ammunition. In the next basket a buzz-cut blond with a snazzy dress uniform hawks over homeland security, an American eagle shield on his arm, and a machine gun, pistol, Bowie knife, two grenades, truncheon, and handcuffs at the ready.

2. From *US News & World Report*, November 30, 2007:

BAGHDAD, Iraq—In the back of an armored Stryker vehicle bound for one of Baghdad's more volatile neighborhoods, the US military is transporting what is perhaps the most controversial weapon in its counterinsurgency arsenal today: civilian anthropologists...

The Army began training social science recruits for Iraq this year, christening the teams with a classic military appellation—human terrain system. The name may not be an attention-grabber, but the mission has been: The teams act as advisers to brigades, mapping the relationships (human terrain in military parlance) of the power players and the local people...

US military officials are working hard to reassure wary academics that this is no covert intelligence operation, charges that have been fueled by ads placed by contractors on job-search sites (including some that specialize in intelligence careers), which request that HTS [human terrain system] applicants have experience in the "intelligence arena"...

Over dinner at FOB [forward operating base] Falcon, [HTS team members]... decompress... They know, they say, that anthropologists in war zones are easy targets for satire—even within the brigade. They joke about soldiers coming to them to ask about the cultural implications of, say, killing an entire town—a bit of admittedly dark humor, they are quick to add...

3. From *Newsweek*, October 16, 2007:

HADITHA, Iraq—The Marines know how to get psyched up for a big fight. In November 2004, before the Battle of Fallujah, the Third Battalion, First Marines, better known as the "3/1" or "Thundering Third," held a chariot race. Horses had been confiscated from suspected insurgents, and charioteers were urged to go all-out. The men of Kilo Company—honored to be first into the city on the day of the battle—wore togas and cardboard helmets, and hoisted a shield emblazoned with a large K. As speakers blasted a heavy-metal song, "Cum On Feel the Noize," the warriors of Kilo Company carried a home-made mace, and a ball-and-chain studded with M-16 bullets. A company captain intoned a line from a scene in the movie *Gladiator*, in which the Romans prepare to slaughter the barbarians: "What you do here echoes in eternity"...

In Haditha, the Marines of Kilo Company sometimes handed out candy to kids but mostly patrolled about in Humvees... It is not clear exactly what happened in Haditha on the morning of Nov. 19. One Marine and 24 Iraqis died, that much is certain. Local survivors say Americans on a rampage massacred their neighbors in cold blood...

****

Taken together, these accounts lead us to pose a question: Do they fairly represent our society and fellow citizens, or do they represent dark, surreal exceptions to a saner American reality? Speaking in only a slightly different context, historian Howard Zinn once noted that "isolated oddities can, on investigation, prove to be deviations from an otherwise healthy set of circumstances. Or they may turn out to be small symptoms of a more generalized malady."

Deviations or symptoms? This collection of essays addresses the dimensions of this problem.

A Militarized Society

The US is a deeply militarized society in ways that are often invisible to those living inside it. In the words of Andrew Bacevich, a self-described conservative Vietnam vet who now teaches international relations at Boston University:

> Today as never before in their history Americans are enthralled with military power. The global military supremacy that the United States presently enjoys—and is bent on perpetuating—has become central to our national identity... The nation's arsenal of high-tech weaponry and the soldiers who employ that arsenal have come to signify who we are and what we stand for.

Some numbers begin to tell the story. Over half of all the weapons sold in the world are manufactured by American corporations, and the US accounts for 48 percent of all military spending in the world. Put another way, the US spends almost as much on the military as every other country in the world combined. (Despite this, or maybe because of it, Americans seem to be more afraid of being attacked than citizens of any other industrialized society).

It was a Republican President and war hero, Dwight Eisenhower, who said (in 1953) that "every gun that is made, every warship launched, every rocket fired signifies, in the final sense, a theft from those who hunger and are not fed, those who are cold and are not clothed." So what to make of a country that spends 58 percent of its discretionary budget on the military? This is more than it spends on education, environmental protection, transportation, veterans benefits, housing, job training, agriculture, energy and economic development combined.

Where other industrialized countries have welfare states, the US has a warfare state. While over 40 million Americans have no health insurance, the US has poured over $100 billion into a ballistic missile defense system that most scientific experts believe will not work. In lieu of developing a national economic policy, politicians, feeding what Seymour Melman referred to as America's "permanent war economy," devote themselves to increasing government funding for local military bases and defense contractors. The result is a military budget that, in constant dollars, is the highest it has been since World War II.

A society whose political economy is so militarized inevitably militarizes its relationships with other nations. Thus the US maintains about a thousand military bases abroad and it has 1.68 million men and women in uniform versus only 6,000 foreign service officers and 2,000 USAID employees. As David

Kilcullen observes, the US has more military band players than foreign service officers! Meanwhile the military's share of foreign aid activities rose from 6 to 22 percent between 2002 and 2005, and American diplomats have complained that they are increasingly being supplanted by military attaches abroad. According to a 2006 *New York Times* article, a report written by the Republican staff of the Senate Foreign Relations Committee expressed unease that "some embassies have effectively become command posts with military personnel all but supplanting the role of ambassadors in conducting American foreign policy."

And then there is what Jackie Orr has called "the militarization of inner space." We are everywhere surrounded by, and thus internalize, what Catherine Lutz in this volume calls a "permanent state of war readiness." Every day our government tells us what level of alert we are on for terrorist attack. The video games our teenagers play are so saturated with military scenarios that the military now sees them as a recruitment and training tool. Hollywood's movie industry—the industry that most defines the United States in foreign eyes—churns out films, from *Rambo* to *Saving Private Ryan*, that glamorize soldiers, projecting them as archetypal American men whose struggles define the quest to come of age and find meaning in life. As David Robb reveals in his book *Operation Hollywood*, the filmmakers behind such films routinely submit their scripts to the Pentagon and rewrite them, in exchange for access to military hardware and military locations, if the Pentagon objects to particular scenes. *Top Gun*, *The Right Stuff*, and *Iron Man* are all examples of blockbusters rewritten at the Pentagon's behest. Just in case teenagers are insufficiently brainwashed by their constant marinating in this pro-military mass culture, under the No Child Left Behind Act high schools lose all their federal funding if they do not give the names, addresses, and phone numbers of their students to military recruiters. Meanwhile universities that refuse to exempt military recruiters from their non-discrimination policies can lose their federal funding.

There are now 25 million veterans in the US, making veterans and their families a larger electoral constituency than either trade unions or the Christian right. No wonder a Democratic Presidential candidate seeking to end the war in Iraq chose to introduce himself at his party's convention with the words "Lieutenant Kerry reporting for duty." What C. Wright Mills called a "military metaphysic" has become so powerful in our society that politicians often feel they can only speak against a particular war from within this metaphysic.

And now in our discipline's analogue to John Kerry reporting for war duty as a lieutenant (flanked protectively, so he thought, by Swift Boat veterans), we have anthropologists who say they oppose the Iraq war, but are helping the Marines write their counterinsurgency manual and deploy to Iraq in uniform, supposedly to save lives. American anthropology substantially demilitarized after World War II, unlike most other disciplines in the social sciences, and it looked carefully and courageously into what was expected of it in the Vietnam years and decided as a discipline to turn its back. Now, in the context of a society more deeply militarized than ever, a new generation of anthropologists must decide whether they will be washed along with the tide of militarism or will maintain their discipline's status as a largely demilitarized academic zone.

Culture as a Weapon

Today's push to weaponize anthropology is but the latest wave of a prolonged series of campaigns to use anthropology to meet the needs of warfare and occupation. Anthropology has been of service to military planners and colonial managers since its formal disciplinary conception in the nineteenth century when European colonialism and American notions of manifest destiny drew upon ethnographic knowledge and theory to inform the management of conquered or occupied peoples.

There were scattered uses of anthropology in the First World War, but the most significant impact for the field came at the war's end, when Franz Boas published a letter in *The Nation* denouncing four unnamed anthropologists who had used a pretext of conducting archaeological fieldwork as a cover for Office of Naval Intelligence espionage missions in Mexico and other nations along the Gulf of Mexico. Boas wrote that anyone "who uses science as a cover for political spying... prostitutes science in an unpardonable way and forfeits the right to be classed as a scientist." A few weeks after Boas' letter was published he was censured by the American Anthropological Association (AAA), not because he was mistaken in his accusations, but because he had spoken against individuals who were seen as doing their patriotic wartime duty.

It was during World War II that American anthropology was formally organized to meet military and intelligence needs in innovative ways. Over half of America's anthropologists contributed their disciplinary skills to the war, doing a variety of things ranging from generating language and culture pocket guides issued to GIs being shipped out to battle theaters, to conducting covert operations in regions they knew from pre-war fieldwork. Some of this work was classroom-based and though those taught were soldiers or intelligence operatives, their work was not fundamentally different from the academic duties of most academics. Anthropologists worked on war problems related to refugees, domestic war production problems, intelligence analysis and propaganda development. Samuel Lothrop, one of the archaeologists criticized by Boas for spying during the First World War, reprised his role of spy while pretending to conduct archaeological fieldwork in Peru. About a dozen anthropologists were hired by the War Relocation Authority, and assisted in the internment of Japanese-American citizens living in the American west. A few heroically stood up for the mistreated internees, but most simply managed the internees—manipulating their environment to keep them complacent and manageable.

Some of this work raised ethical questions for these anthropologists, while others were unconcerned. Anthropologist Laura Thompson worried about what would become of anthropology as it was so easily harnessed for warfare. In 1944, Thompson wrote that she was worried that the war had turned anthropologists and other social scientists into nothing more than "technicians for hire to the highest bidder." That such misgivings appeared during what most Americans saw as a "good war" helps us understand why so many anthropologists have expressed serious concerns about the abuses of anthropology in the more problematic wars that followed.

While professional ethics codes did not exist until after World War II (in fact all professional ethics codes derive from the Nuremberg Code), many of anthropology's contributions to the war were essentially educational and fell permissibly within the sort of ethical parameters that would later be clarified in the ethics codes of the Society for Applied Anthropology and the AAA, though some uses of anthropology during the war were clearly outside these parameters.

Anthropologists' involvements in the Vietnam War brought increased scrutiny to anthropology's roles in warfare and to general questions of anthropological ethics. As public opinion turned against the war, movements within the AAA converged to adopt a code of ethics that clarified not only responsibilities to research participants, and requirements of disclosure, but declared the impropriety of secret research and of reports that populations being studied could not access. However, these secrecy prohibitions were removed during the Reagan years as the result of applied anthropologists' successful lobbying to have the ethics code permit their work with proprietary data.

The history of anthropology's episodic militarization finds three recurring fundamental ethical problems: secrecy, harm, and manipulation. The non-transparent nature of much military and intelligence work violates conventional academic norms on the open circulation of knowledge as well as undermining anthropology's traditional commitments to research participants. Ethical commitments to do no harm can be compromised when anthropologists study "enemy" cultures and ethnographic data is being provided to military and intelligence agencies. Applying anthropology to manipulate cultures raises serious ethical issues;

the inherently non-democratic nature of counterinsurgency's manipulation of other cultures is directly opposed to normalized understandings of acceptable anthropological practices.

Applying anthropology in warfare raises political and ethical questions. The political questions regarding the justifications or causes of a war are generally evaluated by individuals; the ethical issues of particular uses of anthropology in warfare are debated at a collective level. As American politicians press political agendas with military power, anthropologists must be concerned about the political ends to which the discipline is being put. The Network of Concerned Anthropologists is centrally concerned with both the ethical issues raised when anthropology is used for militarized ends, as well as the political ends to which these programs are being applied.

Invasions don't need anthropologists, but occupations do; historically, American military and intelligence agencies have recognized the need for specific forms of cultural knowledge as a foundation of effective counterinsurgency. Today, the US military is seeking to relearn these lessons from anthropologists. Do anthropologists want to get involved with enabling unjust occupations? Increasingly, calls for assistance will be framed as "humanitarian aid" in zones of collapse triggered by American invasion and occupation. Anthropologists must consider their roles in these cycles of violence and "relief."

Counterinsurgency: A Brief History

In its broadest sense, counterinsurgency refers to the methods by which a dominant power imposes its will upon a subordinate population. Specifically, it refers to the elimination of an uprising against a government.

Insurgency, *insurgent*, *surge*, and *insurrection* all share the same etymological roots—they entered the English language from the Latin word *surgere*, to rise up, circa 1500. However, *counterinsurgency* has a much more recent history. Its origins date back to 1960, when the US Department of Defense issued a classified Special Forces manual, *Counterinsurgency Operations*. In 1962, the RAND Corporation (a federally funded research center) convened a famous symposium entitled "Counterinsurgency" in which military men discussed techniques and theories.

Even though the term counterinsurgency is of recent provenance, the US Army has long been involved in such work. A historian notes that the Carlisle Barracks (site of the US Army War College) was constructed in central Pennsylvania in the mid-1700s "for the purpose of instructing British and Provincial troops in counterinsurgency, which back then meant fighting Indians." Much later, counterinsurgency became a label for a type of warfare that had existed for centuries. It has gone by a range of names including "imperial policing," "counter-guerrilla" or "counter-revolutionary" operations, "small wars," and (beginning in the 1970s) "low-intensity" conflict.

In general, the goal of counterinsurgents, imperial police, counter-guerrillas, counter-revolutionaries, and those engaging in "small wars" was (and is) the suppression or annihilation of revolutionary movements in occupied territories.

During the Peninsular War (1807-1814), irregular Spanish fighters ambushed, sabotaged, and harassed Napoleon's occupying army through "hit-and-run" operations. The *guerrilla* (literally "small war") tactics greatly debilitated French forces as Spanish insurgents demoralized and weakened them over time—a vital element in Britain's final triumph.

When this style of "guerrilla warfare" erupted in European colonies in the late 19th and early 20th centuries, it was dealt with ruthlessly: in the Cuban War of Independence (1895-1898) and the Philippine Revolution (1896-1898) against Spain; in the French colonies, where Joseph Gallieni developed strategies for quelling insurrectionists in Indochina (1892-1896); and in the Anglo-Boer War (1899-1902), an insurgency that followed Britain's attempted annexation of Boer lands. (British Colonel C.E. Caldwell wrote the book *Small Wars*—still considered one of the foundational texts on counter-guerrilla operations—based partly upon his experience in the latter.)

After the US defeated Spain in the Spanish-American War (1898), Philippine rebels demanded sovereignty, but President McKinley refused. Instead, the US became involved in a bloody war to counter insurgents, who continued using guerrilla tactics but ultimately lost the conflict. Many American troops involved in the "Philippine Insurrection" (1898-1902) had experience in fighting the Indian Wars. In letters and diaries they described their opponents as uncivilized savages.

When the British engaged insurgents militarily they often assigned creative names to the conflicts: the

"Troubles" in Ireland (1920s), "Arab disturbances" in Palestine (1920s and 1930s), and Mau Mau and Malayan "Emergencies" (1950s). Today counterinsurgency theorists describe the British campaign against communist guerrillas in Malaya as a successful counterinsurgency in which success depended upon separating the civilian population from the disputed territory by forcibly relocating them to "new villages." (This was essentially a response to Mao Zedong's guerrilla tactics, which called for revolutionary insurgents to "move through the people like a fish moves through water.") British techniques were later emulated by the US military in Vietnam under the auspices of the "strategic hamlets" program.

French troops engaged in counter-revolutionary warfare in Indochina (1946-1954) and Algeria (1954-1962). They failed miserably against the Vietnamese, who were led by Ho Chi Minh, and though they appeared to have defeated the Algerian National Liberation Front militarily by 1958, the counterinsurgency wars had taken a tremendous human toll in Algeria and France, and domestic and international public opinion soon turned in favor of Algerian independence.

In US-occupied Haiti (1915-1934), Nicaragua (1909-1933), and the Dominican Republic (1916-1924), Marines Corps officers honed counter-guerrilla tactics and eventually codified their techniques in the *Small Wars Manual*, published in 1940—a precursor to subsequent counterinsurgency manuals. In the "small wars" (which were anything but small from the point of view of the Haitians, Nicaraguans, and Dominicans living under the yoke of American occupation), US troops were deployed to safeguard the economic interests of US investors. However, not all agreed with these policies. For example, Major General Smedley Butler later regretted his participation:

> I spent 33 years and four months in active military service and during that period I spent most of my time as a high class muscle man for Big Business, for Wall Street, and the bankers. In short, I was a racketeer, a gangster for capitalism. I helped make Mexico and especially Tampico safe for American oil interests in 1914. I helped make Haiti and Cuba a decent place for the National City Bank boys to collect revenues in. I helped in the raping of half a dozen Central American republics for the benefit of Wall Street. I helped purify Nicaragua for the International Banking House of Brown Brothers in 1902-1912. I brought light to the Dominican Republic for the American sugar interests in 1916. I helped make Honduras right for the American fruit companies in 1903... Looking back on it, I might have given Al Capone a few hints. The best he could do was to operate his racket in three districts. I operated on three continents.

Butler's comments underscore the criminality and brutality of counterinsurgency warfare, and its connection to the expansion of corporate capitalism. Although its contemporary proponents emphasize "winning hearts and minds," counterinsurgency is always a dirty business—a means of crushing resistance to imperialism. During the Philippine Insurrection, US troops used the "water cure" (waterboarding) and other forms of torture, and inflicted indiscriminate violence upon women and children. In the end, "pacification" of the Philippines was at the cost of approximately

200,000 Filipino deaths. By 1961, French officer Roger Trinquier included torture among the techniques of *Modern Warfare* against insurgents:

> If the prisoner gives the information requested, the examination is quickly terminated; if not, specialists must force his secret from him. Then, as a soldier, he must face the suffering, and perhaps the death, he has heretofore managed to avoid.

Today, many US military planners still consider Trinquier among the most influential counterinsurgency theorists.

* * * *

Insurgencies stir ambivalent sentiments in our country. Consider these words, written by analysts at the RAND Corporation in 1970:

> Insurgency is a subject that is especially difficult for Americans to view with dispassion... We are, or conceive ourselves to be, an insurgent people originating in a tradition of rebellion against inequitable, onerous, and illegitimate authority... The American attitude of sympathy and attachment toward an insurgent cause is not inconsistent with a readiness to react, even overreact, with massive military force against those insurgencies we have wittingly or unwittingly become committed to oppose for reasons of supposedly realistic international politics... The combination of a Jeffersonian heritage with contemporary international politics disposes us to oscillate between sympathy and identification with an insurgency on the one hand and impassioned and self-righteous hostility on the other.

Perhaps it is for this reason that the two most heralded strategies employed by the US in Iraq are a "surge" and—simultaneously—"counter-surge" (or counterinsurgency).

Conceptually the ideas are in tension. But the reality is clear enough. On the ground, Iraq's counterinsurgents are men like Adnan Thabit, who commands 5,000 troops and is described admiringly as "The Godfather" by some US officers. Thabit has a close relationship with American advisers, notably James Steele, who is among the US military's top counterinsurgency experts.

According to *The New York Times Magazine*, Steele—who was appointed by General David Petraeus (then commander of US forces in Iraq) to work with Thabit and other commanders—gained notoriety in El Salvador in the 1980s, where a US-supported right-wing government battled a leftist insurgency. More than 70,000 people died, mostly at the hands of right-wing forces. A team of 55 US Special Forces advisers, led by Jim Steele, trained battalions accused of human rights violations. A kinder, gentler counterinsurgency is a myth and a fantasy.

About NCA and the Volume

The Network of Concerned Anthropologists was formed in the summer of 2007 when eleven like-minded anthropologists began corresponding and searching for ways to express concerns over recent efforts to militarize anthropology. We decided to take collective action and produce a statement of our objections to developing trends in the militarization of anthropology. This statement was loosely modeled on a document circulated by physicists, computer scientists and engineers in the mid-1980s opposing Ronald Reagan's Strategic Defense Initiative and pledging to decline funding to participate in it. The two physicists who originated that statement, David Wright and Lisbeth Gronlund, went on to work for the Union of Concerned Scientists; hence the name we chose for ourselves. Our own statement, worked out over a period of weeks through intense email exchanges, clarified our shared objections to military and intelligence agencies' uses of anthropology in the present political context. We circulated the statement among colleagues and posted it on our website[*], collecting over 1,000 signatures from like-minded anthropologists and other scholars. (The website was mysteriously taken down on the final day of the 2007 American Anthropological Association meetings, but was soon restored). Like other historical movements within American anthropology (such as the Anthropologists for Radical Political Action of the early 1970s), we work for change within professional organizations such as the AAA or the American Association of University Professors.

---

[*] http://concerned.anthropologists.googlepages.com/

Not all members of the Network share the same critiques of military uses of anthropology. Some members of the Network are opposed to all forms of military employment by anthropologists, while others limit their critiques to specific relationships, particularly those involving secrecy or those that risk betraying standard relationships that emerge when we do ethnography. These differences and our ability to make common cause over larger issues is a fundamental strength of the Network.

The Network generates public critiques of new developments and policy proposals. We do not oppose engagement with military and civilian policy makers; we want to expand public debates to include informed critiques that use scholarship and political and ethical critiques to move towards better policies and practices. We strive for a form of public anthropology that engages the public and policy makers on topics that include the ethics and efficacy of the Human Terrain System, the Minerva Consortium, counterinsurgency, or abuses of anthropological research.

Although some bloggers accused the Network of being a small group of elitist tenured radicals telling less fortunate and more junior anthropologists what they should and should not do, in fact only six of the eleven founders have tenure. Two others are assistant professors, one is an adjunct, one a graduate student and another a professional editor. It bears mentioning also that one of our founders works in Canada and another in Europe, and that one is a military veteran. Again, in our diversity lies our strength.

When Marshall Sahlins generously approached the Network's Steering Committee and asked if we had anything appropriate for his Prickly Paradigm Press series, we decided to produce this volume as a resource to other scholars interested in learning more about the issues raised by the military harvesting of knowledge from academic research, to describe fundamental problems with counterinsurgency, and to offer examples of how needed critiques can be produced at a grassroots level.

Any author royalties generated from this volume are being divided between the Network of Concerned Anthropologists and Iraq Veterans Against the War. We encourage our readers, once they have read this book, to join with us and to seek their own ways, as anthropologists and as citizens, in which to undo the damage done to Iraq and Afghanistan by US military intervention and to join us in the struggle to contest the militarization of anthropology.

Part I
Counter-histories of Militarism

Chapter 1
The Military Normal: Feeling at Home with Counterinsurgency in the United States
Catherine Lutz

Here were two typical US media moments in July 2008. On Fox News, Brit Hume opened his interview with two advocates of "victory" in Iraq—pundits Charles Krauthammer and Fred Barnes—and with a video clip of George Bush explaining rising US casualties in Afghanistan. "One reason why there have been more deaths [recently]," the President declared, "is because our troops are taking the fight to a tough enemy. They don't like our presence there because they don't like Americans denying safe haven."

This clip was followed by one of Admiral Mike Mullen, Chairman of the Joint Chiefs of Staff, reminding Americans, "We all need to be patient. As we have seen in Iraq, counterinsurgency warfare takes time and a certain level of commitment. It takes flexibility."

Over on radio, NPR's *Fresh Air* program featured an hour-long discussion with Lieutenant Colonel John Nagl, a recently retired Army officer who led tank assaults in the First and Second Gulf Wars, one of the co-authors of the *US Army/Marine Corps Counterinsurgency Field Manual*, and now working with the ascendant, new Democratic leaning think tank, the Center for a New American Security. Interviewer Terry Gross's gentle questions generally went like this one:

> Ethics become very complicated because it's hard to tell who's a friend and who's the enemy. And that must make it hard to tell, too, when it's appropriate to fire and when it's not. Can you tell us about a difficult judgment call you had to make about whether or not to fire on individuals or on a crowd?

Prevailing mainstream media discussions of counterinsurgency wars in Iraq and Afghanistan have this restricted kind of range, focusing on how the wars are being fought, or should be fought—with what tactics, for how long, and with what level of "success." The pundits, with the populace in tow, debate whether the military is stretched too thin, well-enough resourced (or not), or in need of tens of thousands more troops to do the job. They debate whether the Bush administration lied about the reasons for going into Iraq, but not whether the nation should have been there or in Afghanistan under any circumstances. They debate timetables for bringing the troops home, not plans for accountability for illegalities of war and torture.

They do not ask whether the US should have history's most lethal and offensively postured military, one with soldiers garrisoned in approximately 1,000 bases around the world, waging wars covert and overt in numbers of countries, and with annual costs to citizens of $1.2 trillion and an arsenal of unparalleled sophistication in ways to destroy people and things. They treat it as a "no brainer" that the security of the United States grows when the military budget or the size of the army grows, and that it is sensible for the federal government to spend more on the military than on protecting the environment, educating children, building transport systems, developing energy sources, agriculture, and job skills in the populace, and getting people into housing—combined. In these discussions, it goes without saying that the military serves the nation and/or the world as a whole, "policing" it for the common good. These stories assume that above all we live in a world of threat and risk, of enemies and allies, and of nation and state rather than global and human interests as operative values. They assume that all civilians addressed by those media outlets are American citizens who are happy to pay the Pentagon's bills and who want nothing but victory or honorable withdrawal from fights around the world. The dominant media narrative suggests the values associated with the military as an institution—obedience, loyalty, duty, honor, conformity—should be the primary values of the civilian world as well; it assumes that soldiering builds character more than nursing does.

Although polls show the majority of Americans in 2008 wanted US troops to return home from Iraq, the reasons pundits and populace most often gave for this did not question those basic assumptions: they argued the fight was elsewhere, or that the military should be put in a more defensive rather than interventionist stance. Everyone is inside the consensus that makes an enormous military normal and acceptable: the most left-leaning think tanks tend to, at most, propose cutting military spending by 10 percent, pulling it back to levels it was at just a few years ago. Whether conservative or centrist, libertarian or liberal, the TV, radio and Internet sites from which the great majority of Americans get their news share these elements of a foundational narrative about the military, war, empire, and the world "out there," a world whose voices are rarely heard.

For expert comment on security, the media go to generals and civilian Pentagon employees who have made their careers preparing to make war far more than to diplomats, humanitarian officials, or civilians about to deal with the catastrophic consequences of war. Many of these experts, *The New York Times* reported in 2008, have been operating on specific Pentagon instructions as to message, with access to Defense Department power brokers and, in some cases, money as compensation. These Pentagon mouthpieces are but the tip of a very large iceberg of money invested in convincing the American public to support a government and economy on a permanent war footing, a point to which I return in a moment.

We can call all of this—the massive investments in war and in the public relations of war, and the assorted beliefs that sustain them all—"the military normal."

A Brief History of the Rise of the Military Normal

In his important 1956 book, *The Power Elite*, C. Wright Mills wrote that a military mindset had already, relatively early in the era of permanent war, taken deep hold among the public. He termed this now taken-for-granted world of assumptions a "military metaphysics," or "the cast of mind that defines international reality as basically military." He argued that "The publicists of the military ascendancy need not really work to indoctrinate with this metaphysics those who count: they have already accepted it." He noted that what he called "crackpot realists" had come to rule the day, garnering virtually unlimited power to use a "military definition of reality" to pursue power, status, profit, and their vision of the US role in the world. Mills' argument that the US operated primarily at the relatively unified behest of a power elite of government, corporate and military elites, whose center was in military institutions, contrasts with a common pluralist notion of how things work. In that latter scheme, democratic checks and balances exist at many levels, protecting against the concentration of power in any segment of society, including the military. Civilian control of the military is said to be one of those checks.

Instead, Mills argued, a strong unanimity of thinking about war and foreign policy had emerged among civilian and military elites as a result of their shared identity as people different from, and in fact superior to, the average citizen. Their agreement on military matters comes from having walked in each others' shoes: large numbers of them regularly move between government, corporations, and the military. Former high-ranking officers end up as well-paid employees of the companies they once bought weapons from, or they become members of Congress, granted special status as security policy experts or as appropriations committee members. And they share a metaphysic because institutional recruitment filters ensure new members enter only on accepting the military definition of reality, as evidenced by the progressive community organizer who ran for president in 2008 and came to argue for attacking Pakistan if necessary and increasing the size of forces deployed in various war zones.

The ascendance of the military came about only relatively recently in US history. While the US, as a state, was born through violence—Indian Wars, the Revolution, and slave repression being the most important forms that violence took—it was founded on a suspicion of standing armies, and with civilian leadership ensured by Constitutional frameworks. Military leaders had relatively limited powers as a result: the public saw the military as a burden in peacetime and at best very occasionally necessary. Government-run armories and shipyards provided limited incentives for politicians and the business sector to argue for increased military spending. Middle class families were reluctant to send their children into a military they saw as a virtual cesspool of vices.

World War II brought unprecedented levels of spending and coordination between private enterprise and the Federal government and massive conscription of young men into a 10-million-person force. Money, story lines, and cultural capital were spread around for all to share. The war set in motion a process of militarization that has waxed and waned but never truly flagged in the seven decades since it began. The war gave the US a far-flung empire of bases, economic prerogatives, and cultural influence but militarization resulted not just from the attempt to sustain these but from the incentives of all of the institutions and groups who benefited from a large military budget. Not only weapons makers but companies like Procter & Gamble and the Disney Corporation came to enjoy and rely on immense military contracts. US universities were drawn up in a concerted government campaign to put much of the nation's scientific talent and university training at the disposal of the military, to the point where 45 percent of all computer science graduate students with federal support get it from the Pentagon, and 25 percent of all scientists and engineers work on military projects. The military-industrial-Congressional-media-entertainment-university complex is a massively entangled system.

The end of the Cold War resulted in a dip in military spending but it was no steeper than fluctuations in the past many decades. In that post-Cold War period as well, the military normal began to cover an even more expansive swath of the national horizon, and by two routes. First, strategic planners argued that the US military, now absent any competitors, should aim to exert "full spectrum dominance" over untoward events anywhere, not just fight and win wars in which US national security would be deemed to be at risk. As a result, the number of US wars grew rapidly, with 33 open interventions in just the period from 1990 to 2008. Second, claims were made that the military was now better suited than any other institution to conduct all aspects of US foreign policy from war-making to humanitarian rescue to development aid. The non-military components of the State Department and US aid budgets and functions shrank in direct result, with the regular Pentagon budget now 30 times their combined size. The Pentagon share of official development assistance rose from 6 percent of the whole in 2002 to 22 percent and rising in 2005. Foreign aid and diplomacy now often run, not ambassador to ambassador but military to military, strengthening those armies overseas as well. Many embassy staff have also been complaining about the recent proliferation of military attachés at US embassies around the world.

The rapid growth of the Pentagon budget and functions, particularly in the post 9/11 period, has included takeover of other agencies' intelligence functions, even within the security sector. The Pentagon's intelligence budget is now $60 billion, dwarfing the CIA's $4 to 5 billion. The military normal suggests that this buying and spying, whoever does it, harvests security. The difficulty of even following the changes and scale of funding has also made it difficult to see how inconsequential are the merely multimillion-dollar initiatives, like the Human Terrain System or Minerva, meant to ramp up the cultural intelligence used by the military.

Violence and the American Self-image

Despite the huge military budget, the frequent interventions overseas, and the morphing of foreign policy into military policy, Americans have been convinced that their nation is peace-loving (even when some relish the idea that no one trifles with the US without swift retribution). In its long-standing war narrative, the US fights reluctantly, rarely, and defensively, as Tom Engelhardt has so eloquently noted in his *The End of Victory Culture*. It is only when attacked that Americans rise to defend themselves. So it was a crucial and effective technique for the Bush administration to falsely link Iraq and the 9/11 attackers, to garner support for a war that was offensive in most senses of that term. Every invasion, in fact, has been portrayed this way—from the Dominican Republic in 1965 to Vietnam in the 1960s to Panama in 1989.

Remarkably, military and civilian leaders have even been successful in convincing people that the military rarely engages in killing. Military recruitment ads and official pronouncements from Pentagon, White House, and Congress suggest as much. As Elaine Scarry has pointed out, these discussions replace the broken bodies at war's center with war's supposed purpose, such as toppling a tyrant or freeing hostages. The preferential option for a view of the military as a defensive and innocent force for good in the world helps explain why US soldiers participating in invasions and armed attacks are most commonly described as "putting themselves in harm's way." Even images of troops being killed are now censored, with *The New York Times* counting just six realistic images of dead US soldiers over the entire war and across all media outlets. The work of moral hygiene that each day comes out of the mainstream media, and the Pentagon public relations and recruiting offices which feed them material, have made the war that most Americans know an almost non-physical imposition of will on others.

The military normal is sustained in part through this sense of innocence, a sense bestowed and maintained in two ways. The first involves the fictionalization of American war history via the ascendance of the Hollywood definition of reality, many of whose war films have had official Pentagon support. Besides the large new harvest of Iraq and Afghanistan veterans who know what combat looks like, most students I teach about war have learned what they already know about it through the gloss of film and TV. They begin their questions and discussions with the military normal that those movies help reproduce, even if the message is sometimes the resigned notion that "war is hell."

The second involves the heavy censorship and cleaning up of actual wars' reality for public media consumption. CBS could receive a call from Dick Cheney in 2004 telling them not to publish the Abu Ghraib story, and the station would sit on the information for several weeks, emerging with it only after a leak of the official Taguba report to journalist Seymour Hersh made public knowledge inevitable. The embed system for reporters has successfully kept critical journalists from sources and outlets for their stories, and exerted great pressure on those who do embed to report warmly on the men and women in uniform, an additional incentive being provided by the fact that these people help keep the reporter alive. This was all further strengthened by a concerted campaign to portray the US military, as Andrew Bacevich has noted, as having invented a new, humane, highly targeted form of warfare. It was one based in smart weaponry and new strategy that decapitates demagogues rather than assaults a nation, one that sends bombs through the eye of a needle to wreak vengeance only on "the bad guys." This vision of a new, even more civilized American Way of War predated the recent celebration of the rise of General Petraeus's smarter, less kinetic brand of counterinsurgency warfare, but very much sets the context for the enthusiastic reception of Petraeus, the new *Counterinsurgency Field Manual*, and the idea of gathering academics to provide cultural knowledge to the military.

So the manual was launched in 2007, a year when, by conservative estimate, the US military killed 713 Iraqi citizens and was involved in firefights where 1000 more were killed. The military normal is increasingly oriented around the idea of the exception—the civilian death as an exception, America as the exceptional nation, and the exception from rules called for by states of emergency, an emergency now decades long. It is guided by the spirit of a sign that Stan Goff, a Special Forces veteran, reports has hung in a Fort Bragg training area to encourage the sense of initiative desired in unconventional warriors: "Rule #1. There are no rules. Rule #2. Follow Rule #1." To be above the law is to be within the military normal.

This vision of the US military is also sustained by having two versions of every document that guide military activity—one more a document of civilization and the other of barbarism. So the *Counterinsurgency Field Manual* published by the University of Chicago Press has a doppelganger manual, the *Foreign Internal Defense Tactics Techniques and Procedures for Special Forces* (published in 1994 and 2004). The public relations version of past US military action used by the Chicago version of the counterinsurgency manual sits in front of the well-recorded history of actually existing US counterinsurgencies in places like Vietnam and El Salvador where the techniques of torture, assassination, and massive killing of civilians were in common and approved use.

The Counterinsurgency Campaign in the United States

It has often been said that modern warfare is centered in public relations more than weaponry, that the side that commands the story told about the fight—its rationale, justness, victims, and heroes—will win the war. Less often recognized is how much war and the military normal have depended on public relations campaigns at home, among the American public who must be convinced to continue to supply people to fight and money to buy weapons. From World War II's Office of War Information to the $20 million contract to monitor US and Middle Eastern media coverage of news from Iraq in order to "promote more positive coverage," reproducing support for war has required heavy lifting and significant investments.

    The Pentagon has an annual budget of almost $3 billion for advertising and recruitment of new troops, a massive investment shaping domestic opinion. The GAO reported that, in the three years from 2003 to 2005, the Department of Defense spent $1.1 billion hiring advertising, public relations, and media firms to do the work of convincing the public that the war is important, going well, and requires new recruits and new dollars. The domestic propaganda is directed through the Pentagon's Soldiers News Service, Speakers Service, and other efforts to place news stories and advertising in US media. The military publishes hundreds of its own newspapers and magazines on bases around the world from the *Bavarian News* at US bases in Germany to the *Desert Voice Newspaper* in Kuwait to Fort Hood's *Sentinel* in Texas. It also has its own radio and television stations and a massive network of websites with "news," as one site puts it, on "the purpose and impact of Defense-wide programs," a number of which appear at first look to be civilian sites. The most effective part of the campaign focuses on soldier morale, recruitment, and public opinion simultaneously by encouraging civilians to respect and express gratitude to soldiers for what they do. This is affective labor that many find easy to do, especially when it is in exchange for not having to go to war or send one's children into the military.

This is itself a domestic counterinsurgency campaign, ongoing since World War II, but especially intensified with the resistance that has emerged to the war in and occupation of Iraq. The weapons of this campaign are the ideas articulated by powerful individuals in government and media, on the endless repeat that all marketers know is key to success. It centers on controlling what questions get asked—Is the surge working? Is the Army large enough? Are Human Terrain Teams reducing casualties among civilians? Are our wounded veterans getting adequate care?—more than the answers. And the American public is a more important long-term target than "the Muslim world" or even the general population in Iraq and Afghanistan, because they theoretically control the purse strings and quite literally control whether they put their bodies or their sons' or daughters' bodies in uniform.

The military normal is constructed by a variety of factors, including the impact of years of advertising (only becoming ubiquitous in the media since the institution of the All Volunteer Force in 1973) at a level that not only brings in the requisite number of recruits each year, but convinces its other target audience, the

American public in general, that the military is a reasonable and respectable institution, and that it makes our very way of life possible. In other words, without the military, we would not have the right to free speech or the other democratic freedoms we enjoy. No attention is paid to the fact that these freedoms were ensured by legions of civilians campaigning against a state and a military that had come down on the side of their opposite, or that the military has frequently been used to deny those rights to others, as in Chile, Iran or Guatemala. What also plays into the normalization and veneration of the military is the cumulative effect of the nation's 25 million veterans, many of whom are organized in powerful groups that act politically on military and foreign policy questions.

Threats to the Military Normal

The Counterinsurgency Field Manual is not just a cultural artifact of the exotic tribe at the Pentagon, or something created by a few individuals within the Pentagon or prompted by the Bush administration's particular failures and challenges in Iraq and Afghanistan. It is, instead, an artifact of the American whole. While its details might seem exotic, its foundational premises have been these assumptions about the centrality and necessity of military force as the core of US state functions. And its successful launch as a message about what the military can accomplish—coercion without blood—depended on this decades-long work to control the messages that the American people receive and the beliefs they hold about the innocence and high civilizational goals of the US military. It is the outcome of an entire political economy centered on making war and preparing to make war.

How can counter-counterinsurgency hope to work in such a militarized environment?

It can begin by identifying and challenging the pillars of belief and the streams of profit that support business as usual within the military normal. It would be twinned to campaigns for media democracy (given that the military has been able to count on war cheerleading from the merged and acquired corporate media) and for a university system unhooked from its addiction to Pentagon research money and focused on researching how the military-industrial-media-educational system actually works, what its effects are domestically and overseas. It would take some heart from the fact that so much work of public relations must go into trying to hide the war system and its effects from people at home and around the world. Billion-dollar campaigns of domestic counterinsurgency suggest that the hearts and minds at stake are at home in the United States.

Chapter 2
Militarizing Knowledge
Hugh Gusterson

A Cautionary Tale from Physics

At the beginning of World War II, some of the world's most brilliant physicists saw that it might be possible to build an atomic bomb and became afraid that Nazi Germany was attempting to do just that. They persuaded Albert Einstein to write to Franklin Delano Roosevelt, lobbying for an emergency bomb-building program. The result was the Manhattan Project. Although the physicists of the Manhattan Project were initially impelled by horror at the prospect of Hitler getting the Bomb first, they soon became intoxicated by the technical challenges of bomb research, and their work only intensified in winter 1944 when it became plain that Hitler's bomb project under Heisenberg had failed. By the spring of 1945, when Germany was defeated and the original rationale for the Manhattan Project completely disappeared, only one scientist had left Los Alamos—Joseph Rotblat (who would later go on to win the Nobel peace prize for his disarmament work with the Pugwash organization). And so it was that the Manhattan Project, conceived as an attempt to outpace the Nazi atomic bomb project in defense of Western civilization, metamorphosed into an offensive project targeting Japan. Its scientists produced the bombs that obliterated Hiroshima and Nagasaki, killing over 200,000 people, most of them civilians.

Many of the physicists who had worked on the Manhattan Project decided not to do nuclear weapons work after the war. Some, such as Philip Morrison and Joseph Rotblat, became active in the cause of disarmament. In a story that is well told in fine new biographies by Kai Bird and Martin Sherwin and by Priscilla McMillan, J. Robert Oppenheimer, the father of the atomic bomb, tried to use his position as an insider and adviser to the US government to head off attempts to develop a hydrogen bomb. However, the US government disregarded the unanimous advice of its Strategic Advisory Committee (chaired by Oppenheimer) not to build the hydrogen bomb. Instead, the government followed the advice of Oppenheimer's arch-rival, Edward Teller, embarked on a crash program to develop the H-bomb and opened a second nuclear weapons lab at Lawrence Livermore. In 1954, at McCarthyism's high ebb, Oppenheimer was punished

Militarizing Knowledge

for his dissent by being stripped of his security clearance. He died a broken man a decade later.

By the 1950s, what had started as a defensive maneuver to stop the Nazis from acquiring an atomic monopoly produced a terrifying nuclear arms race between the Americans and the Soviets. By the early 1980s, this led to a world with over 50,000 nuclear weapons—enough to kill every man, woman and child on the planet several times over. Even now, almost two decades after the end of the Cold War, it is proving hard to dismantle the entrenched infrastructure of nuclear weaponry: The US still retains thousands of nuclear weapons and, at over $6.5 billion per year, spends more (in constant dollars) on nuclear weapons research and development now than it did during the Cold War.

One set of lessons we might draw from this story has to do with unforeseen consequences: scientists embarked on a defensive program to defend the West against a German atomic bomb threat and instead ended up spearheading an offensive program that destroyed two foreign cities and ignited a race to accumulate destructive stockpiles of unimaginable power. Another set of lessons has to do with the difficulty of finding a reverse gear for processes of militarization: even someone as powerful as J. Robert Oppenheimer found his advice disregarded and eventually saw his career destroyed when he sought to stop and reverse the nuclearization of international relations that he had helped begin. Processes of militarization are, in other words, hard to control: they escape the grasp of their originators.

But I want here to focus on the effects of these processes on physics itself—a discipline that became deeply militarized during World War II and the cold war. In his forthcoming book *American Physics and the Cold War Bubble*, the historian of science David Kaiser explores what happened to physics during World War II and the Cold War. First, resources poured into the discipline, facilitating the construction of expensive reactors and cyclotrons and underwriting the expansion of physics departments at universities all over the United States. A field that had, in the 1930s, graduated fewer than 100 PhDs a year was producing 500 a year by 1955 and a staggering 1,500 a year by 1970. But although physics experienced expansion in terms of resources and enrollments, in other ways it experienced a kind of narrowing or shrinkage. For one thing, like an agricultural system that has shifted to export-oriented monoculture, it became deeply dependent on defense funding. Kaiser reports that 98 percent of all federal funding for academic physics came from the Department of Defense and the Atomic Energy Commission (AEC) by 1954. Three-quarters of the students whose research was funded by the Atomic Energy Commission went on to work for the AEC upon graduating. Meanwhile, despite the explosion of resources, there was a narrowing of research interests: a discipline that had been characterized by multiple research foci in the 1930s was, by the 1950s, largely focused on nuclear physics and solid-state physics—two areas of great interest to the military and the defense industry. "In no other country did the horizon of young physicists' research topics constrict so dramatically," says Kaiser. Kaiser also observes that American physicists largely dropped the interest in deeper philosophical issues that had characterized pre-war physics and continued to prevail in European physics departments. Instead, they adopted a pragmatism that resonated with the outlook of their sponsors. Perhaps this is what the historian Stuart Leslie was getting at in his book *The Cold War and American Science* when he identified one consequence of the Cold War as "our scientific community's diminished capacity to comprehend and manipulate the world for other than military ends."

By the time I did my own field research among nuclear weapons scientists in the 1980s, the nuclear weapons laboratories at Los Alamos and Lawrence Livermore had become the largest employers of physicists in the country. Between them, they employed six percent of all US physicists. According to some estimates, as many as 40 percent of physicists in the 1980s were doing some kind of military work. Some of the weapons scientists I interviewed freely admitted that weapons work had not been their first choice, but that the job market afforded them few other options. The political economy of physics served as a gigantic funnel that tended to channel new physicists toward careers in military research. For lack of funding, they were drawn away from careers developing alternative energy technologies, new means of public transportation, or medical devices.

The Cold War University

Variants of the story of physics can be told for many other academic disciplines. In her history of Stanford University, *Creating the Cold War University*, Rebecca Lowen shows how Stanford's administration was able to leverage military funding to expand the university and build up particular departments, but at a cost. Lowen's book repeats the story Kaiser tells for physics: an explosion of resources accompanied by a narrowing of vision and a deformation of departments to suit new patrons. In Stanford's case, the administration was quite ruthless in pushing aside those who criticized the new directions set by military sponsors or whose research was of no interest to the military. The university administration was in part driven by its interest in recovering as much money as possible in overhead on federal grants so that the university could continue to grow. Fields such as ecology, political philosophy, and field geology were of little interest to the military, and although researchers in these fields were doing good work, these fields began to be de-emphasized in the curriculum, in hiring decisions, and in appointments of department chairs. Fields such as electrical engineering and particle physics were of great interest to the military and they were marked for growth. Just as Kaiser observed that postwar American physics succumbed to a comfortable pragmatism, so Lowen says political science at Stanford began to eschew the deeper philosophical questions and engagement with centuries of political theory that had predominated before World War II in favor of a pragmatic search for workable empirical generalizations that might guide policy-makers. Lowen observes in particular that ethics was downgraded in the curriculum of Stanford's political science department. The communications department, responding to the national security state's interest in measuring and shaping public opinion, shifted toward survey research and polling. At a time when the US national security state was concerned about the apparent ability of communist regimes to manipulate public opinion and inculcate obedience, the psychology department at Stanford began to emphasize research on polling, market research, obedience to authority, and mind control. Meanwhile, as the superpowers vied with each other for influence and supremacy across the Third World, Area Studies Centers (deeply reliant on military funding) appeared not just at Stanford, but at universities all over the country. The Russian Research Center at Harvard and the Center for International Studies at MIT were among the most prominent.

Furthermore, according to Lowen, the balance between the humanities and the sciences shifted dramatically in favor of the latter. In a relatively short period of time the military had helped to reshape the topography of knowledge and reorder the hierarchies within and between disciplines at Stanford. At the same time, as the administration became increasingly focused on bringing in large research grants, undergraduate education—formerly the central mission of the university—was de-emphasized in favor of research. Education became increasingly marginal to the mission of the country's leading educational institutions.

Meanwhile, in many fields military funding had introduced a schism within the university between those with and without secret clearances. Those with clearances were invited to special conferences, given access to extra funding, and their students were allowed to work on classified PhD dissertations—a profound violation of academia's traditional ethic of openness. A spectacular example of the way military secrecy created two parallel universes in academia is given by the Harvard geologist Eric Siever in an essay in the book *The Cold War and the University*. During the Cold War the military funded oceanographers to map the ocean floor so that US submarines would have detailed knowledge of the terrain in which they were operating. This exercise produced articles and maps that circulated freely in the open academic world—but these maps, we now know, were not quite right. The US military did not want the maps it had funded to be shared, via the open literature, with the Soviets, and so they initially classified some of the research and then, when some oceanographers protested, "the Navy agreed to publish the maps if [they] were incorrectly plotted." As Siever dryly observes, the Soviets had almost certainly procured good maps from their own oceanographers, "so probably the only sufferers were scientists who were not in the word-of-mouth network and privy to the falsification." Siever's story dramatizes the potentially corrosive effects of secrecy on the academic world: not only were the geologists and oceanographers who were willing to do military work better funded than their colleagues, not only did they have access to maps and knowledge (that would prove important in the development of plate tectonics and other aspects of geological theory) denied their colleagues, but they became complicit in the deliberate injection of misleading knowledge into the open literature. Of course, the deliberate falsification of knowledge is usually a firing offense in academia, but in this instance the military turned it into a patriotic duty.

Anthropology

We know from the work of David Price in particular the effect of World War II and the Cold War on anthropology. Many anthropologists did war work during World War II, but most saw this work as a response to an international emergency rather than as integral to the unfolding of their careers, and anthropology was more substantially demobilized at the end of World War II than was physics. The work in question ranged from Ruth Benedict's celebrated study of Japanese culture at a distance (eventually published as *The Chrysanthemum and the Sword*) to advising insurgency campaigns in the Asian theater and helping to run the Japanese-American internment camps in the US.

Although some anthropologists did work on behalf of the national security state in the late 1940s and the 1950s, the Cold War acted upon anthropology more through a kind of repressive censorship than (as with physics) a positive reshaping of the discipline in response to military funding. These were years when anthropologists on the left were often subject to FBI surveillance, found it hard to get jobs and promotions, and even emigrated. Meanwhile, as was the case with other humanities and social sciences in these years, anthropologists largely avoided any engagement with Marxist theory except for the watered-down, depoliticized form of Marxism one finds in the work of evolutionist anthropologists such as Leslie White. As David Price points out, we will never know what work might have been done but was not by anthropologists whose careers were wrecked by McCarthyism or, more subtly, by anthropologists who feared for their careers if they followed their muse.

In the 1960s, as the US was drawn deeper into war in Vietnam and began to anticipate counterinsurgency campaigns elsewhere in the Third World as well, the national security state began to take a new interest in anthropology. Project Camelot in 1964 was a lavishly funded initiative to mobilize anthropologists and other social scientists to investigate the origins of peasant radicalism and insurgency and devise strategies to preempt, contain, and repress revolutionary movements. Its proponents saw Camelot as an enlightened attempt to reduce the use of military firepower by using social science to forestall the emergence of insurgencies in the first place, but the project excited controversy and outrage, especially in Latin America, when the Norwegian researcher Johann Galtung outed Project Camelot to the public. In the ensuing firestorm, the Pentagon was forced to cancel the project. Many anthropologists reported that the aftershocks of Project Camelot harmed their research, especially in Latin America, for years afterwards as potential academic collaborators and other interlocutors wondered if they were really working for the CIA.

At the same time, revelations that some anthropologists were secretly consulting for the Pentagon in a Southeast Asian village study brought the issue to a boil within anthropology, and the 1968 meeting of the American Anthropological Association was marked by intense and divisive debate over the propriety of anthropologists aiding a war that was seen by many as an unjust war of occupation in which the U.S. military was terrorizing civilians. In 1971, this debate resulted in the adoption of an ethics code ("Principles of

Professional Practice") that used strong language to condemn anthropological participation in secret research and affirm anthropologists' primary obligation not to sponsors, nor to their own governments, but to those they study. "In research, anthropologists' paramount responsibility is to those they study. When there is a conflict of interest, these individuals must come first. Anthropologists must do everything in their power to protect the physical, social, and psychological welfare and to honor the dignity and privacy of those studied," the code began. It went on to affirm that "in accordance with the Association's general position on clandestine and secret research, no reports should be provided to sponsors that are not also available to the general public and, where practicable, to the population studied."

I believe that the American Anthropological Association took this strong stand in part because the kind of people who are drawn toward anthropology are often those with an instinctive sympathy for the underdog and an ability to see their own culture from a distance rather than through red, white, and blue colored filters. At the same time, there was also a certain measure of disciplinary self-interest here: anthropologists can only do research as long as there are people ("informants") willing to talk to them. If they are suspected of working for the CIA or other organs of the national security state, the ranks of potential interlocutors will thin out, and the most interesting research subjects may be the first to disappear. It is partly because of anthropology's unique research method, which involves forging deep bonds of trust with human beings in order to learn about their world, that anthropology took a stronger position against military research than any other social science. Thus in the 1980s, the CIA and other national security agencies would openly recruit at the annual meeting of the American Political Science Association and in its newsletter, and this was hardly a source of controversy within political science, but the same agencies kept away from the American Anthropological Association and its newsletter. While some individual anthropologists worked for military and intelligence agencies in the 1980s and 1990s, the events of the 1960s and 1970s had established the open spaces of the Anthropology Association as off limits to the national security state.

The Future Horizon

Harvard President James Conant described World War II as "the physicists' war." If Defense Secretary Robert Gates has his way, the so-called "war on terror" will be the anthropologists' war. Where his predecessor Donald Rumsfeld thought that countries in the Middle East could be subdued by a combination of American firepower, high technology and superior intelligence, under Gates the Pentagon is taking a cultural turn. From General Petraeus down, military officials are talking about the importance of understanding "human terrain," adversary culture, native languages and so on. Within anthropology itself, the most outspoken proponent of this cultural turn has been Montgomery McFate. Writing in the journal *Joint Force Quarterly* in 2005, McFate said that:

the ongoing insurgency in Iraq has served as a wakeup call to the military that adversary culture matters... The more unconventional the adversary, and the further from Western cultural norms, the more we need to understand the society and underlying cultural dynamics. To defeat non-Western opponents... we need to improve our capacity to understand foreign cultures... [A] federal initiative is urgently needed to incorporate cultural and social knowledge of adversaries into training, education, planning, intelligence, and operations. Across the board, the national security structure needs to be infused with anthropology, a discipline invented to support warfighting in the tribal zone. Cultural knowledge of adversaries should be considered a national security priority.

The signs of the military's restored interest in anthropology are around us everywhere now: attempts to place recruitment ads for the CIA and Human Terrain teams on the American Anthropological Association website; the Pat Roberts Intelligence Scholars Program (PRISP), offering students fellowships in exchange for a commitment to work for the intelligence community on graduation; articles in *The New York Times*, *The Christian Science Monitor*, and *Newsweek* about anthropologists embedded in Human Terrain Teams in Iraq and Afghanistan; a Yahoo group that largely caters to anthropologists who work or consult for the national security state[*]; and the Minerva Initiative—a push by the Pentagon initially capitalized at $50 million to recruit anthropologists and other social scientists to, inter alia, analyze captured Iraqi documents and research the relations between Islam

---

[*] http://tech.groups.yahoo.com/group/Mil_Ant_Net/

and terror. In many ways repeating the language and ambitions of Camelot, Minerva is not a secret project, and recipients of Minerva funding would be allowed to publish in the open literature. However, they would also be required to attend a summer workshop with Department of Defense officials in Washington DC, where they would presumably be constituted as a sort of brain trust for the national security state. In its call for proposals for Minerva, the Department of Defense says it "seeks to increase the department's intellectual capital in the social sciences and improve its ability to address future challenges and build bridges between the Department and the social science community." In other words, the Pentagon wants to establish for the long term that our community is on call and is part of the military's empire of knowledge.

Anthropologists have, to some degree, pushed back against these developments. The Executive Board of the American Anthropological Association condemned Human Terrain Teams as an invitation to place anthropologists in situations where they would be in grave danger of violating the Association's ethics code; the Association formed a special commission to investigate the relations between anthropology and the military; the President of the Association, Setha Low, wrote to the Office of Management and Budget to suggest that Minerva be moved from the Pentagon to a civilian agency; and, in February 2009, a wide majority of the Association's membership voted to restore the strong, clear language of the 1971 ethics code condemning research that is not shared with its human subjects. (This language had been weakened in 1998 in response to lobbying from some applied anthropologists who would like the freedom to share research findings with corporate sponsors, but withhold them from the human subjects without whom the research could not have been done).

The question for anthropology in our time is this: Will anthropology remain largely outside the orbit of the national security state, or has our turn at last arrived—following in the footsteps of physics, chemistry, engineering, political science, communications and psychology—to transform our discipline in response to initiatives from the Pentagon and intelligence agencies? Overnight, Minerva has established the Pentagon as one of the largest funders of anthropological research in the US today. If anthropologists match the Pentagon's cultural turn with their own military turn, it is clear from the history of other disciplines what we can expect: an infusion of resources at the cost, paradoxically, of a narrowing of research foci and points of view; separate conferences and journals for anthropologists who do security work and a widening gap between those with access to the national security state and those without; curricular changes in anthropology, including the emergence of new masters programs, tailored to the production of analysts for the national security state; divisive inquests (think: *Darkness in El Dorado*) into the questionable behavior of future anthropologists who will be incited by Pentagon funders to cross ethical boundaries before which most of us would halt; increasing problems of access in the field for anthropologists, whose careers will become a kind of collateral damage, increasingly seen by human subjects in potential conflict zones as agents of a foreign hegemony; and the progressive marginalization of those, formerly at the discipline's center of gravity, who refuse to undertake this kind of work.

I invite readers to undertake a thought experiment: think of some of your favorite texts in anthropology and ask yourself if the military would have funded them. For myself, I think of the work on symbols by Victor Turner and Mary Douglas, and imagine a Pentagon funding officer asking what possible utility there could be to the military in a symbolic structuralist theory of the Abominations of Leviticus; I think of the *Writing Culture* volume that transformed anthropology in the 1980s and imagine the funding officer dismissing it as fluff—and dangerous fluff given the volume's meditation on anthropology's prior complicities with colonialism; I think of Paul Farmer's work on the damage wreaked on Haiti by American policies and hear the funding officer muttering, "We don't want to go there"; I think of Emily Martin's pathbreaking work on the human body in biomedicine and suspect the funding officer will say, "Let NIH fund it if they have any money left"; and I think of Philippe Bourgois' riveting ethnography of New York crack dealers—a book that has changed the way many of my undergraduates look at the world around them—and imagine our funding officer asking, "But how is it going to help us in Iraq?"

Just as American physics during the Cold War narrowed its focus and became more pragmatic, less philosophical, so we can imagine that the kind of anthropology funded by the military would be less engaged with international high theory, less literary, less reflexive, less radical, and less engaged with what Michael Fischer and George Marcus famously called *Anthropology as Cultural Critique*. Instead it is more likely to be geographically clustered around areas in the Middle East, Latin America, and Africa of interest to the military, to have a clear payoff for its funders, and to be written in the kind of bullet-point prose one finds in political science.

Many mainstream commentators argue that the United States has reached a point where it must decide whether or not it wants to be an Empire. Similarly anthropology has reached a point where it must decide whether it wants to be the human relations branch of Empire. It is my most profound hope that anthropologists will refuse to transform their discipline into one that uses a rhetorical patina of cross-cultural understanding and harm reduction to mask a project that would understand the other in order to subjugate and control it. This would be a betrayal of our human subjects and of our vocation as interlocutors of the other.

Part II
Countering the Counterinsurgency Manual

Chapter 3
Faking Scholarship: Domestic Propaganda and the Republication of the *Counterinsurgency Field Manual*

David Price

> If I could sum up the book in just a few words, it would be: "Be polite, be professional, be prepared to kill."
>
> —John Nagl, *The Daily Show with Jon Stewart*

Soon after the U.S. Army and Marine Corps published the new *Counterinsurgency Field Manual* in December 2006, the American public was subjected to a well orchestrated publicity campaign designed to convince them that a smart new plan was underway to salvage the lost war in Iraq. In policy circles, the manual became an artifact of hope, signifying the move away from the crude logic of "shock and awe" toward calculations that rifle-toting soldiers can win the hearts and minds of occupied Iraq through a new scholarly appreciation of cultural nuance.

Things were going poorly in Iraq, and the American public was assured that the *Counterinsurgency Field Manual* contained plans for a new intellectually fueled "smart bomb" for victory in Iraq. This contrivance was bolstered in July 2007, when the University of Chicago Press republished the manual in a stylish, olive drab, *faux*-field ready edition, designed to slip into flack jackets or Urban Outfitters accessory bags. The Chicago edition included the original foreword by General David Petraeus and Lieutenant General James Amos, with a new foreword by counterinsurgency expert Lieutenant Colonel John Nagl and introduction by Harvard's Sarah Sewell. Chicago's republication of the *Counterinsurgency Field Manual* spawned a media frenzy, and Nagl became the manual's poster boy, appearing on NPR, ABC News, NBC, and the pages of *The New York Times*, *Newsweek*, and other publications, pitching the manual as the philosophical expression of Petraeus' intellectual strategy for victory in Iraq.

The Pentagon's media pitch claimed the manual was a rare work of applied scholarship, and old Pentagon hands were shuffled forth to sell this new dream of cultural engineering to America. Robert Bateman wrote in the *Chicago Tribune* that it is "probably the most important piece of doctrine written in the past 20 years," crediting this success to the high academic standards and integrity that the Army War College historian, Conrad Crane, brought to the project. Bateman touted Crane's devotion to using an "honest and open peer review" process, and his reliance on a team of top scholars to draft the *Counterinsurgency Field*

Manual. This team included "current or former members of one of the combat branches of the Army or Marine Corps." As well as being combat veterans, "the more interesting aspect of this group was that almost all of them had at least a master's degree, and quite a few could add 'doctor' to their military rank and title as well. At the top of that list is the officer who saw the need for a new doctrine, then-Lt. Gen. David Petraeus, PhD."

The manual's PR campaign was extraordinary. In a *Daily Show* interview, John Nagl hammed it up in uniform with Jon Stewart, but amidst the banter Nagl stayed on mission and described how General Petraeus collected a "team of writers [who] produced the [counterinsurgency] strategy that General Petraeus is implementing in Iraq now." When Jon Stewart commented on the speed with which the manual was produced, Nagl remarked that this was "very fast for an Army field manual; the process usually takes a couple of years;" but for Nagl this still was "not fast enough." The first draft of each chapter was produced in two months before being reworked at an Army conference at Fort Leavenworth, Kansas. The speed with which the *Counterinsurgency Field Manual* was produced should have warned involved academics that corners were being cut, but none of those involved seemed to worry about such problems. The manual's insertion into mainstream American popular culture was part of the military's larger scheme to use willing glossy outlets to convince the American public that new military uses of culture would lead to success in Iraq. While one conservative magazine criticized these efforts (*The American Spectator*), the liberal press (*The New Yorker*, *Elle*, *More*, *Wired*, *Harper's*, etc.) climbed on board, running glossy uncritical profiles of the cultural counterinsurgency's pitchmen in glamorous write-ups portraying this new generation of anthropologists as a brilliant new breed of scholars who could culturally co-opt foreign foes and capture the hearts and minds of those we'd occupy. The willing press pitched the Pentagon's message that top scholars were now using scholarship to prepare America for victory in Iraq.

The American public was assured that in Iraq and Afghanistan the military was implementing the manual's approach to the use of culture as a battlefield weapon. Human Terrain Teams (HTTs) embed anthropologists with troops operating in Iraq and Afghanistan, and the *Counterinsurgency Field Manual* was hailed as the intellectual tool guiding their coming success.

The Secrets of Chapter Three

The heart of the *Counterinsurgency Field Manual* is Chapter Three's discussion of "Intelligence in Counterinsurgency." It introduces basic social science views of elements of culture that underlie the manual's approach to teaching counterinsurgents how to weaponize the indigenous cultural information they encounter in specific theaters of battle. General Petraeus bet that troops working alongside HTTs could apply the manual's principles to stabilize and pacify war-torn Iraq and Afghanistan.

When I read an online copy of the *Counterinsurgency Field Manual* in early 2007, I was unimpressed by its watered-down anthropological explanations, but having researched anthropological contributions to World War II, I was familiar with such oversimplifications. Like any manual, it is written in the dry, detached voice of basic instruction. But when I reread Chapter Three a few months later, I found my eye struggling through a crudely constructed sentence and then suddenly being graced with a flowing line of precise prose:

> A ritual is a stereotyped sequence of activities involving gestures, words, and objects performed to influence supernatural entities or forces on behalf of the actors' goals and interests.

The phrase "stereotyped sequence" leaped off the page. Not only was it out of place, but it sparked a memory. I knew that I'd read these words years ago. With a little searching, I discovered that this unacknowledged line had been taken from a 1972 article written by the anthropologist Victor Turner, who brilliantly wrote that religious ritual is:

> a stereotyped sequence of activities involving gestures, words, and objects, performed in a sequestered place, and designed to influence preternatural entities or forces on behalf of the actors' goals and interests.

The manual simplified Turner's poetic voice, trimming a few big words and substituting "supernatural" for "preternatural." The *Counterinsurgency Field Manual* used no quotation marks, attribution, or citations to signify Turner's authorship of this barely altered line. Having encountered students passing off the work of other scholars as their own, I know that such acts are seldom isolated occurrences; this single kidnapped line of Turner got me wondering if the manual had taken other unattributed passages. With a little searching, I found about 20 passages in Chapter Three showing either direct use of others' passages without quotes, or heavy reliance on unacknowledged source materials.

The numerous instances I found shared a consistent pattern of unacknowledged use. While any author can accidentally drop a quotation mark from a work during the production process, the extent and consistent pattern of this practice in the *Counterinsurgency Field Manual* is more than common editorial carelessness. The cumulative effect of such non-attributions is devastating to the manual's academic integrity, and claims of such integrity are the heart and soul of the Pentagon's claims for the manual—claims that the military hoped to bolster with the republication of the *Counterinsurgency Field Manual* at a top academic press.

The use of unquoted and uncited passages is pervasive throughout this chapter. For example, when the manual's authors wanted to define "society" they simply "borrowed" every word of the definition used by David Newman in his *Sociology* textbook; they lifted their definition of "race" from a 1974 edition of *Encyclopedia Britannica*; and their definition of "culture" was swiped from Fred Plog and Daniel Bates' *Cultural Anthropology* textbook. The manual's definition of "tribe" was purloined from an obscure chapter by Kenneth Brown, and not only is Victor Turner's definition of "ritual" hijacked without attribution, but the manual's definition of "symbols" was a truncated lifting of Turner. Several sections of the *Counterinsurgency Field Manual* are identical to entries in online encyclopedia sources. The manual's authors used an unacknowledged truncated version of Anthony Giddens' definition of "ethnic groups." Max Weber's definition of "power" is taken from *Economy and Society* and used without attribution. And so on. Each of these passages was taken without the use of either quotation marks or any acknowledgment that real scholars had originally written these words.*

Other sections of the manual have unacknowledged borrowings from other sources. Roberto González discovered that the manual's Appendix A was "inspired by T.E. Lawrence, who in 1917 published the piece 'Twenty-seven articles' for *Arab Bulletin*, the intelligence journal of Great Britain's Cairo-based Arab Bureau." González compared several passages of Lawrence with David Kilcullen's Appendix A, and found parallel constructions where paragraphs were reworded but followed set formations between the two texts. González observed that while these parallel constructions can be seen, "Lawrence is never mentioned in the appendix." González shows that Kilcullen's other written work makes a passing reference, "but does not acknowledge the degree to which Lawrence's ideas and style have been influential."

A complicating element of the *Counterinsurgency Field Manual*'s reliance on unattributed sources is that the manual includes a bibliography listing of over 100 sources, yet not a single source I have identified is included. My experience with students trying to pass off the previously published work of others as their own is that they invariably omit citation of the bibliographic sources they copy, so as not to draw attention to them. Even without using bibliographic citations, the manual could have just used quotes and named sources in the same standard journalistic format used

---

* For an earlier version of this critique and examples and sources of these passages see: http://www.counterpunch.org/price10302007.html

in this chapter, but no such attributions were used in these instances.

The inability of Chapter Three's authors to come up with their own basic definitions of such simple sociocultural concepts as "race," "culture," "ritual," or "social structure" not only raises questions about the ethics of the authors, but also furnishes a useful measure of the manual and its authors' weak intellectual foundation. In all, I quickly found over a dozen examples of lifted passages from uncredited sources.

When I published an exposé in October 2007 documenting the extent of the *Counterinsurgency Field Manual*'s plagiarized passages in *CounterPunch*, the military had a variety of responses. Officially, US Army spokesman Major Tom McCuin issued a doublespeak statement declaring a mistakes-were-made-but-the-message-remains-true admission that passages were indeed used in an inappropriate manner. Less officially, a mob of blood-boiling counterinsurgency believers furiously blogged on the *Small Wars* website attacking me, my credentials, and my reputation and discussed plots designed to get me fired from my job—those involved in this plagiarism have suffered no negative consequences, while tenured academics like Ward Churchill on the left are made to suffer for any involvements in any form of plagiarism. Nagl issued a statement claiming that the manual, as military doctrine, did not need footnotes or attributions of any type. Nagl's response skirted the issue of the manual's lifting exact sentences (and of slightly modifying others) and reproducing them in the manual without quotation marks as if the problem were simply one of missing footnotes and citations and not of lifted quotations. Nagl wrote that it was his "understanding that this longstanding practice in doctrine writing is well within the provisions of 'fair use' copyright law." A few military scholars, like historian Lieutenant Colonel Gian Gentile of West Point publicly criticized Lieutenant Colonel Nagl's lame excuses and argued that the academic credibility of the manual had been undermined.

In one sense, the particular details of how the *Counterinsurgency Field Manual* came to reprint the unacknowledged writings of scholars do not matter. If quotation marks and attributions were removed by someone other than the chapter's authors, the end result is the same as if the authors intentionally took this material. The silence on the reproduction of these passages, the lack of any authorial *erratum*, and the failure to add quotation marks even when Chicago Press republished the manual seems to argue against the likelihood of a simple editorial mix-up, but who knows? The ways that the processes producing the manual so easily abused the work of others inform us of larger dynamics in play when scholars and academic presses lend their reputations, and surrender control, to projects mixing academic with military goals.

Criticizing the *Counterinsurgency Field Manual*'s rejection of the most basic of scholarly practices is not (as Nagl later tried to argue) holding it to external standards; it is to hold the manual to its own standards. Nagl later argued that using the unattributed passages of others is acceptable when writing military doctrine. But the preface of the University of Chicago Press's edition of the manual clearly says: "This publication contains copyrighted material. Copyrighted material is identified with footnotes. Other sources are identified in the source notes." According to doctrine's preface, doctrine has footnotes. The instances in which the manual does use quotes and attributions provides one measure of its status as an extrusion of political ideology rather than scholarly labor, as these instances most frequently occur in the context of quoting the apparently sacred words of generals and other military figures—thereby, denoting not only differential levels of respect but different treatment of who may and may not be quoted without attribution.

After my critique was published, the *Small Wars* website posted a document full of citations and quoted passages that purported to be an original draft of one problematic section of the *Counterinsurgency Field Manual*'s third chapter. Even as a draft this document has a lot of problems. While it has an impressive use of footnotes, there remain sentences (often marked with footnotes) that have no quotation marks yet are the words of others. I don't know the provenance of this document, but even if it were the original draft of a chapter that was later altered by unknown citation-and-quote-removing editors, it does not answer basic questions of why chapter three's authors remained silent when the University of Chicago Press republished a work they would have then known to have contained the unacknowledged work of others. If this is what happened, why was no *errata* forthcoming? The mysterious production of this claimed early draft without any explanation solves nothing, and raises more questions than it answers.

The numerous footnotes in this supposed "draft" document do shed more light on the extent of anthropologists whose work was consulted in the production of this chapter; these anthropologists include: Clifford Geertz, E.E. Evans-Pritchard,

Napoleon Chagnon, Raymond Firth, A.R. Radcliffe-Brown, Ralph Linton, Bronislaw Malinowski, and Sherry Ortner. I assume that many of the "draft's" cited non-anthropologist radicals such as C. Wright Mills, Antonio Gramsci, or Pierre Bourdieu, would have been disgusted to see their work used to manipulate public opinion in support of military occupation.

The few published critical examinations of the *Counterinsurgency Field Manual* focus on the text's provenience and philosophical roots. In *The Nation*, Tom Hayden links the manual to the philosophical roots of US Indian Wars, reservation policies, and the Vietnam War's Phoenix Program. In *Anthropology Today*, Roberto González observed that the manual "reads like a manual for indirect colonial rule." (That a press as drenched in "reflexive" critiques of colonialism as Chicago would publish such a manual is an ironic testament to just how depoliticized many of postmodernism's salon-bound critiques have become.) A *New York Times* op-ed by Richard Shweder voiced a stance of relativist inaction from which the travesties of the Human Terrain System could be lightly critiqued while anthropologists were urged not to declare themselves as being "counter-counterinsurgency."

The Politics of Republication

The role of University of Chicago Press in bringing the *Counterinsurgency Field Manual* to a broader audience is curious. That such shoddy scholarship passed so briskly through the Press' editorial processes raises questions concerning Chicago's interest in rushing out this faux academic work. Rushing a book through the production process at an academic press in about half a year's time is a blitzkrieg requiring a serious focus of will. There was more than a casual interest in getting this book to market—whether it was simply a shrewd recognition of market forces or reflected political concerns or commitments. The Press enjoyed robust sales of a hot title (it was proclaimed as one of Amazon.com's top 100 in September 2007); to what extent was this due to large advance orders from the Pentagon itself, for example? This is damaging to the Press' reputation.

To highlight the *Counterinsurgency Field Manual*'s scholarly failures is not to hold it to some over-demanding, external standard of academic integrity. It is important to recognize that claims of academic integrity are the very foundation of the manual's promotional strategy. Somewhere along the line, Petraeus' doctorate became more important than his general's stars, touted by Petraeus' claque in the media as tokening a shift from Bush's "bring 'em on" cowboy shoot-out to a nuanced thinking man's war. University of Chicago Press acquisitions editor, John Tryneski, told me the manual went through a peer review process, but there are unusual dynamics in reviewing a previously published work whose authors are not just unknown (common in the peer review process), but in effect unknowable. Tryneski acknowledged that peer reviewers came from policy and think tank circles. When I asked Tryneski if there had been any internal debate over the decision by the Press to disseminate military doctrine, he said there were some discussions and then, without elaboration, changed the subject, arguing that the Press viewed this publication more along the lines of the republication of a key political document. This might make sense if this was an historic document, not a component of a campaign being waged against the American people by the Pentagon, surging to convince a skeptical American public that Bush hadn't already lost the war in Iraq.

Its republication transformed the manual from an internal document of military doctrine into a public "academic" document designed to convince a weary public that the war of occupation could be won: it is an attempt to legitimize the war by "academizing" it. It is troubling that those scholars who worked on the *Counterinsurgency Field Manual* remained silent about attribution problems when they learned of Chicago's plans to republish the manual. If, as some later claimed, quotation marks and citations had been removed by others in the editorial process after the initial draft was submitted, these contributors should have alerted the University of Chicago Press to this. In at least one case, one of the contributors to manual Chapter Three was notified that Chicago would be republishing the manual, but the Press was not alerted to any of these problems.

That militaries commandeer food, wealth, and resources to serve the needs of war is a basic rule of warfare—as old as war itself. Thucydides, Herodotus and other ancient historians record standard practices of seizing slaves and food to feed armies on the move; and the history of warfare finds similar confiscations to keep armies on their feet. But requirements of modern warfare go far beyond the needs of funds and sustenance; military and intelligence agencies also require *knowledge*, and these agencies are evidently looking to commandeer scholarship in ways not intended by their authors.

Commandeering Scholarship for Dirty Wars

The requisitioning of anthropological knowledge for military applications has occurred in colonial contexts, world wars, and proxy wars. After World War II, Carleton Coon recounted how he produced a 40-page text on Moroccan propaganda for the OSS by taking pages of text straight from his textbook, *Principles of Anthropology*:

> I padded it with enough technical terms to make it ponderous and mysterious, since I had found out in the academic world that people will express much more awe and admiration for something complicated which they do not quite understand than for something simple and clear.

The most egregious known instance of recycling of an anthropological text by the military occurred in 1962, when the US Department of Commerce secretly, and without authorization or permission from the author, translated into English from French the anthropologist Georges Condominas' ethnographic account of Montagnard village life in the central highlands of Vietnam, *Nous Avons Mangé la Forêt*. The Green Berets weaponized the document in the field. The military's uses for this ethnographic knowledge were obvious, as leaders of assassination campaigns tried to hone their skills and learn to target village leaders. For years, neither publisher nor author knew this work had been stolen, translated, and reprinted for military ends. In 1971, Condominas described his anger at this abuse, saying:

Faking Scholarship

> How can one accept, without trembling with rage, that this work, in which I wanted to describe in their human plenitude these men who have so much to teach us about life, should be offered to the technicians of death—of their death!... You will understand my indignation when I tell you that I learned about the "pirating" [of my book] only a few years after having the proof that Srae, whose marriage I described in *Nous Avons Mangé la Forêt*, had been tortured by a sergeant of the Special Forces in the camp of Phii Ko.

Today, anthropologists serving on militarily "embedded" Human Terrain Teams study Iraqis with claims that they are teaching troops how to recognize and protect noncombatants. But as Bryan Bender reports in the *Boston Globe*,

> one Pentagon official likened [Human Terrain System anthropologists] to the Civil Operations and Revolutionary Development Support project during the Vietnam War. That effort helped identify Vietnamese suspected as communists and Viet Cong collaborators; some were later assassinated by the United States.

This chilling revelation may clarify the role that Pentagon officials envision for anthropologists in today's counterinsurgency campaigns.

Militarized Anthropology

There is a real demand within the military and intelligence community for the type of disarticulated and simplistic view of culture found in the *Counterinsurgency Field Manual* not because it is innovative but because, beyond information on specific manners and customs of lands they are occupying, this simplistic view of culture tells them what they already know. This has long been a problem faced by anthropologists working in such confined military settings. My research examining the frustrations and contributions of World War II era anthropologists identifies a recurrent pattern in which anthropologists with knowledge flowing against the bureaucratic precepts of military and intelligence agencies faced often impossible institutional barriers. They faced the choice of either coalescing with ingrained institutional views and advancing within these bureaucracies, or enduring increasing frustrations and marginalized status. Such wartime frustrations led Alexander Leighton to conclude in despair that "the administrator uses social science the way a drunk uses a lamppost, for support rather than illumination." In this sense, the manual's selective abuse of anthropology—which ignores anthropological critiques of colonialism, power, militarization, hegemony, warfare, cultural domination, and globalization—provides the military with just the sort of support, rather than illumination, that they seek. In part, what the military wants from anthropology is to offer basic courses in local manners and local mapping so that they can get on with the job of conquest. The fact that so many military anthropologists appear disengaged from questioning conquest exposes a fundamental problem with military anthropology.

As the occupation of Iraq presents increasing concrete problems for the manual's lofty claims of counterinsurgency, its "authors" and defenders take on an increasingly cult-like devotion to their guiding text, a devotion that even finds some betraying the lost cause of Iraq in an effort to save the manual's sacred doctrine. In a December 24, 2007 interview, Charlie Rose gently questioned Sarah Sewell about ongoing disasters in Iraq; Sewell quickly deserted the war she had been recruited to rationalize in order to save the manual, insisting:

> the surge isn't the field manual; Iraq is not the field manual. And I think many Americans tend to conflate these things at their peril. And I think they risk throwing out the baby with the bathwater. If and when we look back on Iraq, it will not mean that the manual was wrong, it will mean that Iraq had very different problems, starting with the legitimacy of the invasion to begin with.

With this twisted logic, the *Counterinsurgency Field Manual*'s use as an instrument of domestic propaganda comes full-Orwellian-circle, as the public is asked to forget that just months earlier a barrage of media appearances by Lieutenant Colonel Nagl and others had pitched the manual as the intellectual foundation for victory in Iraq.

But those selling the *Counterinsurgency Field Manual* to the American public know full well, as it says in the new *Counterinsurgency Field Manual,* that counterinsurgencies, just like "insurgencies, are not constrained by truth and create propaganda that serves their purposes," and the manual's tactics are embraced by intellectual counterinsurgents battling the American public's wish to abandon the disastrous occupation of Iraq.

Chapter 4
Radical or Reactionary? The Old Wine in the *Counterinsurgency Field Manual*'s New Flask
Greg Feldman

Introduction

It warrants a double-take that Sarah Sewall, Director of Harvard's Carr Center for Human Rights Policy, wrote the introduction to the 2007 University of Chicago Press edition of *The US Army/Marine Corps Counter-insurgency Field Manual*. The concept of human rights requires neutrality with respect to warring parties, whether symmetrical fights between nation-states or asymmetrical ones between a state and a rebel faction. Human rights pertain to individuals as individuals rather than as political actors. Therefore, an endorsement of warfare—counterinsurgency or otherwise—by a human rights expert (or, intriguingly, a human rights *policy* expert) signals a neo-conservative co-optation of a liberal platform. Why? Because counterinsurgency's purpose is to maintain an unequal balance of power between the US and the countries it strives to pacify and incorporate into its sphere of influence, even if it requires less "kinetic force" than "conventional" warfare. Unsurprisingly, supporters like Sewall effectively argue that counterinsurgency is an administrative tool, not a humanitarian one, insofar as it aims for stability not justice.

Counterinsurgency Field Manual dedicates nearly 400 pages—inclusive of appendices, diagrams, charts, and historical vignettes—to explaining how counterinsurgency works. Multidisciplinary in scope, it demands a large role from civilian organizations to help in everything from building roads to establishing sanitation facilities. It also draws on a range of expertise from civil engineering to anthropology to public administration to economic development to population control measures to advanced technological surveillance and warfare. The intended synergistic effect is to integrate occupied populations into a US-led global political economy and to terminate insurgents. Anthropologists are pitched as ideal counterinsurgency contributors given their century-long research experience among people on the margins of Euro-American power and their long-term ethnographic methodology through which they develop relations of trust with the people they study. (For the record, very few have accepted offers to participate in counterinsurgency.)

The counterinsurgency concept traces back to British and French colonial policing measures as its proponents readily acknowledge. Counterinsurgency's aim is two-fold: first, to win legitimacy in the eyes of a skeptical local population by providing material security, thereby robbing insurgents of an endless supply of fresh recruits; and, second, to single out committed anti-American insurgents for clean-up operations by small, flexible, and sophisticated military units. The *Counterinsurgency Field Manual* does not explicitly define an *insurgent*, though it explains that *insurgencies* aim to overthrow the existing social order within the state or to break away and form an autonomous state. The spectrum of examples it provides range from the popularly supported French Revolution (one would imagine that US defense intellectuals would have appreciated this insurgency's goal) to a simple military coup d'état that removes a civilian executive. Sewall, however, elaborates that "Many contemporary insurgent movements are, by Western standards, conservative and regressive—seeking to restore social structures and practices threatened by the modern state (or its failure)." Fashionably relativist, on the one hand, and fully able to stop all counterarguments, on the other.

In Sewall's messianic view, counterinsurgency is a "radical" proposition for the US military establishment as it "challenges much of what is holy about the American way of war... Those who fail to see the manual as radical probably don't understand it, or at least understand what it's up against." Radical measures are needed not only to prod a stodgy military bureaucracy into the 21st century but also to heal a nation as "Americans yearn to understand a world in which old assumptions and advantages no longer seem relevant." According to Sewall, the *Counterinsurgency Field Manual* "directly addresses" national disillusionment lest cynicism disengage Americans from their government and the rest of the world.

In contrast to Sewall's brave new world thesis, sent essay argues that counterinsurgency is anything but radical. It is instead a reactionary move riding in well-worn colonial tracks as it is deployed to create relations of dependency between the US and people living within its sphere of influence in Southern parts of the globe. Like colonialism, the mission assumes that occupied populations uninterested in emulating the occupier are unbefitting of modernity in any case. Like colonialism, the *Counterinsurgency Field Manual* also assumes that those who resist it stand outside of history and are impervious to change rather than thoroughly engaged in the contingencies of global politics. This essay exposes the neo-colonial logic and practice of counterinsurgency in which a commitment to cross-cultural understanding is deployed in service of extending US national interests as understood by US defense intellectuals.

The Colonial Elephant in the Room

Counterinsurgency Field Manual deploys a chilling utilitarian logic despite efforts to cast counterinsurgency as an advanced and enlightened form of warfare that respects the peoples and cultures caught in the crossfire. Sewall argues that counterinsurgency is in the business of guaranteeing peace rather than justice. Peace (the more realistic term "stability" is used later in the same paragraph) is an administrative problem about how to pacify an occupied country so that its people would willingly support US strategic interests. In contrast, justice is a political problem because it is open to multiple interpretations, many of which could challenge US interests. Where Sewall punts on justice in favor of stability is precisely the point where the moral argument for counterinsurgency slides down the slippery slope toward neo-colonialism. She touts counterinsurgency as a solution to the US's need to "secure 'ungoverned spaces'" so that it can stabilize the international order. One is quickly reminded of the blank spots on the old European maps of the "Dark Continent" of Africa. Colonial bureaucrats, missionaries, and speculators could not grasp that indigenous societies were organized through kinship rather than bureaucratic states and so they concluded that these were lands of chaos and barbarism. The mere existence of uncontained difference was too unpredictable to leave alone lest it foil grander geoeconomic and geopolitical plans. Indigenous societies would thus soon be governed through colonial administrations that would later transform into post-colonial states. Similarly, counterinsurgency is deployed to create friendly host governments in areas where states fail (or fail to meet Western expectations) as a means to halt the fragmentation of a currently US-led international order. Counterinsurgency's strategy of respecting local culture bears a striking resemblance to the historic liberal appreciation of colonized peoples' customs while simultaneously pitying them for being trapped in naïveté and tradition. In both cases, the superpower justifies its overseas ventures as a reluctant response to uncivilized forces in the world and concomitantly secures geopolitical advantages for itself in the process.

Counterinsurgency Field Manual's historical examples of counterinsurgency practically make the argument themselves that Machiavellian concerns with stability prevail over the moral problems of an occupation. As a case in point, it provides an 1847 vignette of how Irish insurgents crafted their plan to break down the British occupying army:

> The force of England is entrenched and fortified. You must draw it out of position; break up its mass; break its trained line of march and manoeuvre, its equal step and serried array... nullify its tactic and strategy, as well as its discipline; decompose the science and system of war, and resolve them into their first elements.

These Irish insurgents may not have been nice to the occupying forces, but surely the ongoing potato famine, which British rule greatly exacerbated, might explain some of their dissatisfaction. Nevertheless, we are basically asked to conclude from this example that the British were the "good guys" and that Americans should learn how to suppress the poor, hungry, and marginalized through imperial analogy. The same goes for the French example. The *Counterinsurgency Field Manual* points out that Napoleon made a key mistake in his 1808 defeat of Spain by failing to analyze Spanish history, culture, and "suspicion of foreigners" and to plan for counterinsurgency operations. This error, explains the manual, allowed the Spanish to subsequently drain French resources, thus marking the "beginning of the end for Napoleon." Should we blame the Spanish for being "suspicious" about the arrival of the French army? Should our hearts bleed for Napoleon? Perhaps so, if we see his attempt to overthrow conservative, royal states in favor of universalized, French enlightened rule as a precedent for the globalization of US-style capitalism and democracy. The *Counterinsurgency Field Manual* also cites the more recent French use of torture during the Algerian War of Independence (1954-1962) as a case of counterinsurgency gone wrong that cost France its moral legitimacy. Though wisely highlighting torture as a moral mistake, the manual apparently saw no mistake in the basic idea of French colonialism in North Africa that itself inspired the insurgency. That people could have a fundamental problem with an occupation—even if they cooperate with the occupiers—never enters the counterinsurgency equation. Instead, the occupied are asked to accept the US strategic interest (again, as defined by US defense intellectuals) as inherently good.

Not only does Sewall in *Counterinsurgency Field Manual* consciously avoid these complicated political issues, but she "accept[s] war as necessary," albeit reluctantly, and excuses her capitulation as "consistent with the lengthy tradition of Western moral reasoning about war, embodied in the concept of 'Just War'." While we should not be so naïve as to think wars will evaporate if we wish it so, we should wholeheartedly challenge the assumption that warfare has an existence of its *own*. On the contrary, warfare does not exist a priori and independent of historical contingencies, but rather it is an *effect* of global politics in which we all participate. Failure to account for that point permits defense intellectuals to argue that military missions abroad are only grudging responses to "objective" threats to an otherwise universally-acceptable order. Moreover, self-consciously anchoring counterinsurgency in Western moral philosophy does little to counter the charge that it is simply a tool for US geopolitical strategy. (One must also ask if the Western canon itself is as consistent on the Just War thesis as Sewall suggests.)

In this context, it is difficult to accept Sewall's description of counterinsurgency as a *radical* agenda. In its historical context, "radical" emerged as a term to describe an initiative that pushes hard toward the far left. It grew out of historical conditions in which popular political movements challenged monarchical dynasties, European imperialism, and unfettered 19th century capitalism. In contemporary parlance, the term has been applied to extreme positions on either side of the spectrum or just used toothlessly to promote a clever idea that has not yet caught on. More accurately, counterinsurgency is a *reactionary* initiative, which historically has referred to an agenda that pushes things back to the far right, because it seeks to reinvent the colonial relationship for contemporary US purposes by explicitly continuing the British and French colonial trajectory. What else can one conclude given the *Counterinsurgency Field Manual*'s own stock of historical examples? How then is counterinsurgency supposed to work?

Defining the Area of Operation and Mapping the Human Terrain

Counterinsurgency depends upon a total picture of the physical, social, and cultural environment in which it operates. Methods of mapping the physical environment make only a small appearance in *Counterinsurgency Field Manual*, as the Department of Defense already possesses enormous experience in this regard with satellite technology. Suggestions for establishing cooperative relations with host states feature more heavily as counterinsurgency tries to build US-friendly governments in the blank spaces of the US security policy map. However, since counterinsurgency's end game is winning legitimacy from the undecided local population, much of the *Counterinsurgency Field Manual* is dedicated to helping the reader make sense of non-Western (and formerly colonized) societies and cultures. The manual's third chapter entitled "Intelligence in Counterinsurgency" reads like an introductory textbook to cultural anthropology, particularly the subsection entitled "Define the Operational Environment." That task involves creating "a composite of the conditions, circumstances, and influences that affect the employment of capabilities and bear on the decisions of the commander." The commander must prioritize six "civil considerations," listed as areas, structures, capabilities, organizations, people, and events and summarized in the acronym ASCOPE. The *Counterinsurgency Field Manual* highlights "people" as the most important civil consideration. It therefore breaks this category into six parts to gain a more refined understanding: society, social structure, culture, language, power and authority, and interests. A raft of introductory terms are defined so that field commanders and embedded social scientists can—in positivist fashion—identify the key elements of the local population and grasp the social physics that keeps it in motion.

To be sure, many of the basic anthropological terms are updated, suggesting that the *Counterinsurgency Field Manual* authors are attuned to the politics of representation. For example, "race" is defined as a human group, which defines itself or is defined by others through "innate physical characteristics." By realizing that racial categories are culturally constructed rather than biologically determined, the manual escapes association with 19th century social Darwinism, which underpinned the "scientific" idea that a superior white race could groom inferior races for civilization. "Tribes"—a term that anthropologists have abandoned—are defined as genealogically structured groups, which largely determine the member's rights. To soften the edge, *Counterinsurgency Field Manual* adds the qualifier that tribes are also "*adaptive* social networks organized by extended kinship and descent..." Recognizing anthropology's semantic turn thirty-five years ago with (uncited) echoes of Clifford Geertz, the manual describes "culture" as a "'web of meaning' shared by members of a particular society or group within a society." Pierre Bourdieu's concept (also uncited) of social capital also makes a brief appearance, hinting that military brass are more relaxed about French intellectuals than are US conservatives.

Nevertheless, the superficial invocation of analytical concepts attuned to agency, creativity, and relativism cannot conceal the political ends at which counterinsurgency is aimed: the twin goals of co-opting the local population and identifying insurgents for elimination on the basis of social behavior and cultural practice. Precluding the ominous switch from neutral local civilian to potential insurgent depends upon how well the *Counterinsurgency Field Manual* user can read and ameliorate the population's grievances and identify certain individuals capable of participating in an insurgency. Table 3-1 provides the reader with "Factors to consider when addressing grievance," such as "Would a reasonable person consider the population's grievances valid?" and "What does the host-nation government believe to be the population's grievances?" Table 3-2 lists "Insurgency characteristics and order of battle factors" with the former referring to such traits as "popular support or tolerance," "key leaders and personalities," and "support activities, capabilities, and vulnerabilities" and the latter to such points as "composition," "disposition," "logistics," "personalities," and "other factors." Counterinsurgency researchers use sophisticated models of social network analysis to study and visually display relationships between actors with nodes (representing individual people) and connecting lines (representing the relationship between two people). Dyads are described, for example, as "A is a relative of B"; "A is in the same clan as B"; "A likes and respects B"; "A is very similar (same tattoo, same street, same profession) to B," etc. Individuals with more frequent social contact will appear in a graph with a denser, more clustered image. Over time, analysts should be able to discern "normal" patterns of behavior. If clustered individuals deviate from the norm, then counterinsurgency analysts' level of alarm will rise accordingly. The *Counterinsurgency Field Manual* then explains the social organization of a generic insurgency, fleshed out with tactics used to fight the occupying force.

  The process of what the military calls "mapping the human terrain" to prepare counterinsurgency operations begins with Human Terrain Teams (HTTs). HTTs consist of five experts who are attached to a forward deployed brigade. Members include a team leader (a major or lieutenant colonel), a cultural analyst (an anthropologist/sociologist with competence in GIS), a regional studies analyst (an area studies and language specialist), a human terrain analyst (a primary data researcher), and a human terrain research manager (a liaison with other military and intelligence organizations). Defense Secretary Robert Gates has allocated funding for twenty-five teams to support brigades in Iraq and Afghanistan. (By the summer of 2008, there were at least five in the former and one in the latter.) Teams establish relations with the local population to learn of their cultural habits, social organization, political history as well as their concerns and problems. As Jacob Kipp, Director of the Foreign Military Studies Office at Fort Leavenworth, and his co-authors explain in a 2006 article in *Military Review*, the team also gathers data on "key regional personalities, social structures, links between clans and families, economic issues, public communications, agricultural production, and the like." These data are archived and shared with subsequent units operating in the area. The information is also stored in Mapping Human Terrain software, which transmits it to a larger archive in the Reachback Research Center (RRC) based in Fort Leavenworth, Kansas. RRC employs people with ethnographic experience who combine social network analysis with geospatial analysis of villages and the surrounding physical terrain to help identify insurgents and to develop strategies to eliminate them.

Lest we forget that counterinsurgency is as much (if not more) about eliminating insurgents as it is gaining local legitimacy, HTTs' lethal capacity was illustrated in a February 2007 roundtable called "Precision Engagement—Strategic Context for the Long War." John Wilcox—Assistant Deputy Under Secretary of Defense—argued that mapping the human terrain "*Enables* the entire Kill Chain for the GWOT [global war on terror]." It facilitates the tracking of insurgents in real time with its processed information on the geographic distribution of local social networks. This information would kick start a multi-dimensional fortress of high tech surveillance, communications, and weapons systems, which provide what Wilcox calls an "unblinking eye" over the territory in question. That eye informs a weapons grid projection that enables military officers to select the most effective weapon to kill the target. The point is best illustrated in the military diagram in Figure 1. Here we see how counterinsurgency combines anthropology, technology, and quantitative analysis to develop a total picture of what Sewall called "ungoverned spaces" and to remove individuals deemed suspect according to social network analysis.

The basic problem, of course, with social network analysis is that it cannot in all seriousness account for the flexibility in the roles people perform in

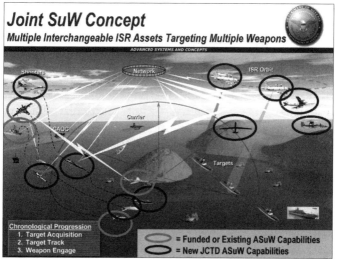

Figure 1.

daily interaction and of the subtle shades of meaning that compose communication between any two people. Regarding the US terrorism trial in July 2008, is the chauffeur to Osama bin Laden automatically a national security threat because he appears in a densely clustered social network diagram with a known terrorist? Under what conditions is the chauffeur working in that role? How accurately can such diagrams identify someone as guilty simply through association when the terms of that association are both difficult to define and change through time? Mistakes in this regard also cost the moral legitimacy of a local population that, reasonably, does not tolerate the death of its kin due to misidentified targets or collateral damage. The point is not even that no analysis should occur if perfect social network analysis is impossible. Rather, it is that the epistemic framework that shapes the units of social analysis is ultimately static, reductionist, and unrealistic. In other words, if counterinsurgency analysts must at some point identify people as either friendly or hostile, then they will miss the subtler intentions and the shades of gray that make up daily social interaction. Ultimately, social relations are qualitative phenomena that cannot be easily quantified, and therefore cannot be operationalized into a military assault, particularly if the point is to secure local legitimacy.

The Lost Lessons of Edward Said

In 1979 Edward Said wrote the influential treatise *Orientalism*, which argued that an apparently innocuous formation of Western knowledge about the East construed colonialism as the natural course of history. As a form of Orientalist writing, the *Counterinsurgency Field Manual* does not recognize what Said and many other post-colonial thinkers have argued: that innocent, liberal-minded social science aimed at appreciating cultural differences actually supports the administration (and thus subordination) of peoples living in geopolitically sensitive places. Orientalism is essentially a Western hegemonic mode of interpreting "other" peoples as morally, technically, culturally, and/or racially inferior that effectively sustains the Western observer's superior position. The very "normality" of this proposition supports an interpretative paradigm shaping everything from academic and literary writing to writings in popular culture to the world of colonial, development, and military policymaking. Said questioned the notion of pure knowledge by demonstrating that such acts of description are inherently political moves rather than innocent exercises of ethnographic description. Orientalism furthermore assumes that the Orientalist (i.e., the Western expert on the East) "is outside the Orient, both as an existential and as a moral fact." This externalization of the Orient justifies "our" involvement in "their" affairs as a reluctant and noble response to problems, which are foreign to "our" history and endemic to "theirs." Through this reasoning, Western efforts to recreate colonized peoples in the image of the West appeared morally clean and free from actual global entanglements.

Said mainly examined how Orientalism functioned in the British and French colonial empires, but there can be no doubt that counterinsurgency is a contemporary Orientalist project. Counterinsurgency relies fundamentally on patronizing colonial discourse as it divides occupied peoples into either modern or regressive and as it depicts those living in geopolitically sensitive areas as in need of a US presence to pacify and develop their countries. Moreover, it marshals not only military experts to its cause but also experts from the liberal arts. This move legitimizes a foreign policy written from the perspective of defense intellectuals. A random example of Orientalism's legacy and counterinsurgency's continuity with European colonial administration is found in a June 19, 2008 letter to the editor of *The Economist* by Hayden Bellenoit, a historian at the US Naval Academy:

> SIR: It should come as little surprise that American forces have adapted to the cultural landscape in Afghanistan quicker than have their British counterparts ("A war of money as well as bullets", May 24th). The British may have pioneered the Great Game, but their political and cultural intelligence about much of north-western India was limited from the onset.
>
> After the 1820s, the British, in a fit of victorious hubris (emboldened by their defeats of Napoleonic France, no less), increasingly cut themselves off from the cultural, linguistic and religious knowledge of the wider Indo-Persian realm. One East India Company official's wistful remark that "beyond the Jumna [river] all is conjecture" could

apply as equally today as it did more than 100 years ago.

It is odd that a nation born of anti-colonial fervor would find its defense intellectuals debating how best to improve upon Britain's ability to play the Great Game. Though its advocates insist on calling counterinsurgency a radical move forward, this interpretation only holds within the confines of that particular community of scholars and policymakers. A broader view suggests that counterinsurgency is a profoundly reactionary project emerging out of a colonial trajectory that privileges stability over justice.

Part III
Countering Counterinsurgency

Chapter 5
Embedded: Information Warfare and the "Human Terrain"
Roberto González

When US-led coalition forces invaded Iraq on March 20, 2003, most Americans received battlefield reports from an unusual source: "embedded" journalists attached to military units. The reporters relied upon American troops and officers as primary sources, often interviewing them in armored personnel carriers, tanks, aircraft carriers, or on military bases. US Marine Corps Lieutenant Colonel Rick Long candidly described the rationale for the embedded journalism program: "Frankly, our job is to win the war. Part of that is information warfare. So we are going to attempt to dominate the information environment."

Within hours after US bombs began dropping over Baghdad as part of a "shock and awe" campaign, the American public was barraged with gripping images of exploding missiles, rumbling Humvees, and menacing fighter planes. Retired generals commented on TV news programs as if they were providing play-by-play coverage of a football game. Images portraying the violence enacted on humans were invisible—the war had been effectively sanitized by TV editors worried about losing access if military officers (or advertisers) didn't like the reports. All the while, viewers were treated to the likes of Fox News reporter Greg Kelly gleefully rolling into Baghdad in an armored vehicle with the US Army's 3rd Infantry Division, CNN's Kyra Phillips reporting from aboard the aircraft carrier *USS Abraham Lincoln*, and dozens of others clad in Kevlar helmets, flak jackets, and camouflage uniforms.

The rise of embedded journalism marked an extraordinary shift in American war correspondence. Rarely had the boundaries between war fighter and reporter been so blurred. Beginning in November 2002—four months before the Iraq invasion—prospective embedded journalists were encouraged to undergo "basic training" in which they were subjected to simulated battlefield conditions. Upon arriving in Iraq, many slept in the same facilities, used the same vehicles, and ate in the same mess halls as the members of their military units. More often than not, the reporters relied upon the troops for their very survival.

A handful of critics critiqued embedded journalism, noting that such reporting amounted to little more than propaganda. By embedding with the US military, they argued, journalists had sacrificed the objectivity necessary to conduct responsible journalism. Veteran journalist and former *New York Times* reporter Gay Talese referred to them as "those correspondents who drive around in tanks and armored personnel carriers, who are spoon-fed what the military gives them... they become mascots for the military."

The few journalists who refused to embed were exposed to severe danger. In several cases, independent (non-embedded) journalists were killed by US troops: ITN reporter Terry Lloyd, cameraman Fred Nerac, and translator Hussein Osman; Reuters cameraman Taras Protsyuk and Telecinco cameraman José Couso; and al-Jazeera reporter Tariq Ayoub were but a few. Many others were killed by Iraqi soldiers and guerrilla fighters. Given the perils facing non-embedded journalists, it was perhaps understandable that so many would choose to attach themselves with military units.

But being an embedded journalist often meant making professional compromises. The very condition of being embedded had a tendency of keeping reporters far away from Iraqi civilians (including killed or injured civilians) and very close to (and empathetic with) US troops. In such an environment, military officials rarely saw the need to censor reports—journalists effectively censored themselves.

Sociologist Andrew Lindner recently completed an analysis of the embedded journalism program. He noted that

> the embedded program proved to be a Pentagon victory because it kept reporters focused on the horrors facing the troops, not the horrors of the civilian war experience... The end result: a communications victory for an administration that hoped to

build support for the war by depicting it as a successful mission with limited cost.

By taking sides, journalists had (wittingly or unwittingly) been used as pawns for waging "information warfare."

Embedding Social Science

Nearly five years later, the Pentagon began a new embedded program: the Human Terrain System, a $190 million initiative designed to embed social scientists with US Army combat brigades. The program consisted of five-person Human Terrain Teams (HTTs) featuring anthropologists and other social scientists attached to combat brigades. Some of the social scientists wore combat fatigues and carried weapons.

HTS began as a relatively small-scale experiment. But in September 2007, Defense Secretary Robert Gates authorized a dramatic expansion of the program. Approximately 25 additional teams were deployed in 2008, with social scientists reportedly earning up to $300,000 for a year-long deployment. According to a former HTS employee, many military commanders don't take the program seriously, but are reluctant to speak out against it because it is "General Petraeus' baby."

Uncritical reports in the corporate media have portrayed HTS as a life-saving initiative that is establishing a kinder, gentler US military presence in Iraq and Afghanistan, although this is completely unsupported by evidence. There is no verifiable data that human terrain teams have saved a single life—American, Afghan, Iraqi, or otherwise. Such reports have all the trappings of a no-holds-barred Pentagon public relations campaign.

The international press has been far less sympathetic. For example, a November 2007 editorial in Mexico's daily newspaper *La Jornada* responded to HTS by noting that "the grotesque cultural mask of counterinsurgent anthropology does not change the brutal nature of an imperialist occupation." Such reactions are perhaps not surprising, given the history of American social scientists' participation in Project Camelot, an ill-fated Pentagon research program designed to employ social scientists for counterinsurgency research (more accurately called counter-revolutionary work) in Latin America.

Even more disturbing is the fact that some military analysts (notably Jacob Kipp, a historian at the US Army's Foreign Military Studies Office [or FMSO] at Fort Leavenworth, Kansas) have openly described HTS as a "CORDS for the 21st century"—a reference to Civil Operations and Revolutionary Development Support, a Vietnam War-era counter-guerrilla initiative. CORDS gave birth to the infamous Phoenix Program, in which South Vietnamese and US agents used intelligence data to target some 26,000 suspected communists for assassination, including many civilians. At the time, CORDS was publicly heralded as a humanitarian effort to win "hearts and minds," while Phoenix simultaneously (and secretly) functioned as its paramilitary arm. This dubious history provides a critical reference point for understanding the potential uses of HTS, even as proponents of the new program use it to whitewash General David Petraeus' counterinsurgency efforts.

Many aspects of HTS raise troubling concerns about the potential abuse of social science by the Pentagon, its subcontractors, and the broader military-industrial complex. These concerns range from the possibility that social science data could be used to target suspected enemies for assassination, to the lack of transparency about the program, to the ethical problems posed by battlefield anthropology. Many are wondering whether wartime collaboration in secretive military projects "prostitutes science in an unpardonable way," as Franz Boas (a founding father of American anthropology) wrote in 1919 in response to anthropologists doing spy work during World War I.

Like the embedded journalism program, the Human Terrain System operates under a peculiar logic: it provides a public relations boost to a failing military occupation. At the same time, it can provide battlefield intelligence directly to military commanders for combat operations. In short, HTS—like the embedded journalism program—functions as a tool for waging "information warfare."

Origins of "Human Terrain"

Human terrain reveals much about the Pentagon world view. The term portrays people as territory to be conquered, as if flesh and blood human beings were a geophysical landscape. Consider the recent words of US Army Lieutenant Colonel Edward Villacres, who leads a human terrain team in Iraq. According to Villacres, his team's goal is to "help the brigade leadership understand the human dimension of the environment that they are working in, just like a map analyst would try to help them understand the bridges, and the rivers, and things like that." This is the language of conquest: it objectifies, dehumanizes, transforms people into things. Like "collateral damage," the phrase vividly illustrates George Orwell's notion of "political language... designed to make lies sound truthful and murder respectable."

Human terrain's reactionary roots can be dated to at least 40 years ago, when the infamous US House Un-American Activities Committee (HUAC) issued a 1968 report singling out the Black Panthers and other militant groups as enemies of the state. The report, entitled *Guerrilla Warfare Advocates in the United States*, included an appendix that stated,

> traditional guerrilla warfare... [is] carried out by irregular forces, which just about always dispose of inferior weapons and logistical support in general, but which possess the ability to seize and retain the initiative through a superior control of the human terrain.

The implication was clear: defusing "guerrilla warfare advocates" such as the Black Panthers would require the US government to wrest control of urban populations.

In the same report, HUAC suggested that urban unrest might require that the President declare an "internal security emergency" which would enable the 1950 Internal Security Act authorizing detention of suspected spies or saboteurs. (Much of the law was repealed in the 1970s, but some elements were restored in the PATRIOT Act.)

After a hiatus, human terrain resurfaced in 2000, when retired US Army Lieutenant Colonel

Ralph Peters wrote an influential article entitled "The Human Terrain of Urban Operations." In it, he argued that it is the "human architecture" of a city, its "human terrain. . .the people, armed and dangerous, watching for exploitable opportunities, or begging to be protected, who will determine the success or failure of the intervention." He described a typology of cities ("hierarchical," "multicultural," and "tribal") and the challenges that each present to military forces operating there: "the center of gravity in urban operations is never a presidential palace or a television studio or a bridge or a barracks. It is always human."

For years, Peters has espoused a bloody version of Samuel Huntington's "clash of civilizations" thesis. In 1997, he predicted that across the world, the US military would need to inflict "a fair amount of killing" for successful conquest:

> There will be no peace... The de facto role of the US armed forces will be to keep the world safe for our economy and open to our cultural assault. To those ends, we will do a fair amount of killing. We are building an information-based military to do that killing... much of our military art will consist in knowing more about the enemy than he knows about himself, manipulating data for effectiveness and efficiency, and denying similar advantages to our opponents.

Since the publication of Peters' human terrain article, literally dozens of intelligence agents, military analysts, Pentagon officials, pundits, and reporters have adopted the term.

In 2006, Jacob Kipp and colleagues from FMSO took the idea a step further by outlining a plan for HTS in the journal *Military Review*. According to

Kipp, US Army Captain Don Smith led its implementation from July 2005 to August 2006 in order to better "understand the people among whom our forces operate as well as the cultural characteristics and propensities of the enemies we now fight."

A $190 Million Corporate Boondoggle

In early 2007, the US Army's Foreign Military Studies Office contracted the British company BAE Systems to begin recruiting social scientists for HTT positions. (Later, MTC Technologies and Wexford Group, a division of CACI, would also recruit team members.)

Proponents of HTS, such as Colonel John Agoglia, insist that the teams "are extremely helpful in terms of giving commanders on the ground an understanding of the cultural patterns of interaction, the nuances of how to interact with those cultural groups." However, Kipp's description of HTS reveals that the program is designed to improve the "gathering" and "operational application" of "local population knowledge."

Kipp and his colleagues describe a process by which this information will be sent to a central database accessible by other US government agencies, including presumably the CIA. Furthermore, "databases will eventually be turned over to the new governments of Iraq and Afghanistan to enable them to more fully exercise sovereignty over their territory." The human terrain teams will supply brigade commanders with "deliverables" including a "user-friendly ethnographic and sociocultural database of the area of operations that can provide the commander data maps showing specific ethnographic or cultural features."

Teams use software developed by the MITRE Corporation called Mapping Human Terrain (MAP-HT). Kipp and his colleagues described MAP-HT as "an automated database and presentation tool that allows teams to gather, store, manipulate, and provide cultural data from hundreds of categories." The Secretary of Defense's 2007 budget justification describes MAP-HT as "a means for commanders and their supporting operations sections to collect data on human terrain, create, store, and disseminate information from this data, and use the resulting information as an element of combat power." It also allocates $4.5 million for MAP-HT between 2007 and 2009.

HTS supporters have unconvincingly argued that such a database would not be used to target Iraqis or Afghans. In a radio interview conducted in the fall of 2007, Montgomery McFate (an HTS architect) stated:

> The intent of the program is not to identify who the bad actors are out there. The military has an entire intelligence apparatus geared and designed to provide that information to them. That is not the information that they need from social scientists.

She claimed that HTT social scientists have "a certain amount of discretion" with data, while providing no evidence that safeguards exist to prevent others from using it against informants. When asked about lack of independent oversight, she answered: "We would like to set up a board of advisors. At the moment, however, this program is proof of concept... [I]t's not a permanent program. It's an experiment."

An experiment without basic ethical safeguards, it might be added, for Kipp notes that

> [to] ensure that any data obtained through the HTS does not become unnecessarily fettered or made inaccessible to the large numbers of Soldiers and civilians routinely involved in stability operations, the information and databases assembled by the HTS will be unclassified.

Theoretically, this data is available for use by the CIA, Special Operations teams, Iraqi police, the Afghan government or military contractors—any of whom might use it for nefarious ends.

Credible accounts have emerged about difficulties plaguing HTS, including ineffective training and gross mismanagement. Former HTT member Zenia Helbig has publicly criticized the program, claiming that during four months of training, there were no discussions about ethical issues such as the potential harm that might befall Iraqis or Afghans. Furthermore, she claims that "HTS' greatest problem is its own desperation. The program is desperate to hire anyone or anything that remotely falls into the category of 'academic,' 'social science,' 'regional expert,' or 'PhD'," which has led to incompetence. According to Helbig, BAE Systems is more concerned with profits, rather than adequate training for team members. Her description of ineptitude and waste characterizing BAE Systems' operations near Fort Leavenworth, Kansas suggests that the company is engaged in war profiteering.

In Their Own Words

On May 7, 2008, Michael Bhatia—a PhD student in political science and HTS member—was tragically killed in a roadside bomb attack in Afghanistan, making him the program's first casualty of war. On June 24, another HTT member and political science graduate student, Nicole Suveges, was killed in a bomb attack in Sadr City, Baghdad. Their deaths underscored the high cost of the program—not only its financial cost, but more importantly its human cost.

Scholars participating in the program are motivated by a wide range of factors. Marcus Griffin, a cultural anthropologist who blogged about his HTS work for nearly a year before suddenly removing all his past posts from the web without any explanation, was driven partly by a sense of patriotism, occasionally signing off with such catchphrases as "freedom is never free." He clearly perceived himself to be a military man—an anthropologist who was "going native" by embedding:

> Going "native" in anthropology is a fairly common strategy to gain a better understanding of the people with whom one is working. I am about a month away from deploying to Baghdad as part of the US Army's new Human Terrain System and have almost gone completely native. How am I doing this? First, I am working out regularly with Lt. Gato. He is showing me how to develop greater strength and endurance... Second, I cut my hair in a high and tight style and look like a drill sergeant... Third, I shot very well with the M9 and M4 last week at the range... Shooting well is important if you are a

> soldier regardless of whether or not your job requires you to carry a weapon. Fourth, I am trying to learn military language with all the acronyms and idioms otherwise alien to university professor such as myself. I actually know what people are saying now half the time... Today among soldiers, I am looking and more often acting just like them... That is what going native is all about: walking in someone else's shoes in order to know what their life is like and therefore why they do what they do.

It is striking that Griffin omits any mention of Iraqis in his description of cultural understanding. For Griffin, "going native" has nothing to do with Iraqis. It means looking, speaking, and acting just like a US soldier.

Others have demonstrated a much greater level of empathy for those living under the yoke of occupation. As the result of my investigations into HTS, I was approached by "Terry" (pseudonym), a former employee of the program. For more than a year, we have maintained email and telephone communication. Terry's warmth, intelligence, and sincerity made a lasting impression. At times, Terry's words reflected an ambivalent and even contradictory position:

> One of the main reasons I was able to build strong rapport and gain more trust from the Iraqis than anyone around me is because I listened to them, learned from them, and at a very deep level respected and admired them. Their lives were as important to me as anyone else's, including my own... More often than not, I was told by Iraqis that I "have a very good heart." They were human to me, and I looked out for their interests as much as I was obligated to work with the military through this program.

Terry later implied that a conflict of interest between the "ethnographic mission" of embedded social scientists and the intelligence needs of military units can create ethical dilemmas:

> One of the issues I would like to address is how important it is for field researchers, or anthropologists, working with the military in a combat environment to maintain integrity to the "ethnographic" mission. HTTs have pressure to prove credibility to the units they serve since we are in the "proof of concept phase," and there may be a tendency for some to respond to anyone who requests information, including intelligence…

Apart from the pressure that some social scientists might feel to respond to such requests, under these circumstances they could easily become unwitting intelligence agents.

What's Human About Human Terrain?

HTS—and HTS data—may perform various functions simultaneously. Images of a "gentler" counterinsurgency might serve as propaganda for US audiences opposed to military operations in Iraq and Afghanistan, propaganda that allows us to fight wars and still feel good about ourselves. Intelligence collected by HTTs will apparently be fed into a database accessible by the CIA, the Iraq police, or the Afghan military for use in targeting suspected insurgents. Agents might employ HTS data to design propaganda campaigns that exploit Iraqi or Afghan fears and vulnerabilities. Each of these examples highlights how HTS can serve as a weapon for "information warfare" and raises grave questions about the appropriateness of embedded social scientists.

If history is any guide, it seems particularly likely that ethnographic intelligence will be used for social control methods reminiscent of those employed by the colonial powers of yesteryear: divide-and-conquer and indirect rule. Consider the words of French commander and colonial administrator Joseph Galliéni (1849-1916), who was an early architect of "pacification" policies in Indochina, French Sudan, and Madagascar in the late 1800s. In a classic statement, he emphasized how ethnographic intelligence could facilitate a divide-and-conquer strategy:

> It is the study of the races who inhabit a region which determines the political organization to be imposed and the means to be employed for its pacification. An officer who succeeds in drawing a sufficiently exact ethnographic map of the territory he commands, has almost reached its complete pacification, soon followed by the organization which suits him best... If there are habits and customs to respect, there are also rivalries which we have to untangle and utilize to our profit, by opposing the ones to the others, and by basing ourselves on the ones in order to defeat the others.

Ethnography can quickly become a "martial art" under these conditions, as noted by anthropologist Oscar Salemink. French colonial officials in Indochina ultimately used ethnographic intelligence to forge counterinsurgency tactics that were "both politically untenable and riddled with contradictions that doomed it to political failures." Despite these failures, HTS seeks to employ similar tactics today.

Given this history, perhaps it is not surprising that in October 2007 the Executive Board of the American Anthropological Association—the largest professional anthropology organization in the US—issued a statement calling HTS "an unacceptable application of anthropological expertise." Five months later, the Society for Applied Anthropology approved a motion expressing "grave concerns about the potential harmful use of social science knowledge and skills in the HTS project."

In the future, historians may question why some social scientists—who over the past century developed the modern culture concept, critiqued Western ethnocentrism in its various guises, and invented the teach-in—enlisted as embedded specialists in an open-ended war of dubious legality. They might wonder why some began harvesting data on Iraqis and Afghans as a preferred method of practical "real-world" engagement. They might ask why, at a time when majorities in the US, Iraq, and Afghanistan wanted a withdrawal of US troops, social scientists supported an occupation resulting in the deaths of hundreds of thousands of civilians.

Counterinsurgency support stands to violate relationships of trust and openness with the people with whom social scientists work. If embedded social scientists are bound by "operations security" or other forms of secrecy, they are not free to share the results of their work with local people. Such work threatens the well being and integrity of all field-based social research—and more importantly, the safety of Iraqi and Afghan civilians. Serving the short-term interests of military and intelligence agencies and contractors is a reckless approach for social scientists to take, for scorched earth research makes it impossible for future investigators to establish the trust necessary for establishing rapport with research participants.

At the height of the Cold War, C. Wright Mills cautioned social scientists about the perils of succumbing to "the bureaucratic ethos. Its use has been mainly in and for non-democratic areas of society—a military establishment, a corporation." He was concerned about the rapid transformation of scientists into mere technicians, lacking any sense of social responsibility for their actions. As those prosecuting the "war on terror" attempt to draw social scientists into ill-conceived operations like HTS, we should reaffirm our democratic values, our professional autonomy, and our social responsibility by refusing to participate.

Chapter 6
Counter AFRICOM
Catherine Besteman

In a televised interview in December 2007, Charlie Rose asked Montgomery McFate for an example of where the lessons contained in the US Army's new counterinsurgency field manual might be applied next. McFate, an advocate for a military embrace of ethnographic intelligence in counterinsurgency campaigns and coauthor of the new manual, responded, "Well, we have something new coming down the pipeline, which is AFRICOM." McFate's comment highlights the expectation that AFRICOM, the newest unified US Combatant Command, would pioneer within the military a new "soft power" approach oriented toward "non-kinetic" force and humanitarian objectives.

(There are currently ten US Unified Combatant Commands, each of which consists of at least two branches of the armed forces and is led by a top ranking military official who reports directly to the Secretary of Defense. The commands have broadly defined missions and are the mechanism through which military intervention is conducted.)

In a January 2008 *Military Review* article, Sean McFate, a former paratrooper who worked with the private military contractor DynCorp International to train Liberian soldiers, echoed his wife: "AFRICOM is a post-Cold War experiment that radically rethinks security in the early 21st century based on peace-building lessons learned since the fall of the Berlin Wall." In his March 2008 testimony to the House Armed Services Committee, AFRICOM's first Commander, General William "Kip" Ward, echoed the McFates:

> AFRICOM is pioneering a new way for a Unified Command to fulfill its role in supporting the security interests of our nation. From inception, AFRICOM was intended to be a different kind of command designed to address the changing security challenges confronting the US in the 21st century.

AFRICOM, the newest US Combatant Command, was created under EUCOM in October 2007 and will be launched as a separate command by September 30, 2008. Whereas the continent was formerly divided between three different US Unified Combatant Commands (European Command, Central Command, and Pacific Command), AFRICOM encompasses all African countries except Egypt under one US command. In addition to signaling a major reorganization of the US command structure and a new US focus specifically on Africa, President Bush and AFRICOM transitional team leaders publicized AFRICOM as a dramatically new paradigm within the Department of Defense. "Unlike traditional Unified Commands, Africa Command will focus on war prevention rather than war-fighting," reports the AFRICOM website. But is the US military the best vehicle for war prevention? Is AFRICOM good news for war prevention in Africa?

Reviewing the history of US military/security engagements with Africa suggests an ominous future for AFRICOM. AFRICOM is the newest component of a profoundly damaging and destructive US foreign policy based on militarism and militarization. Defining security in Africa as something that can be created through increasing the flow of weapons, the power of armies and the presence of US troops is contrary to common sense as well as historical experience. If military security is the objective, why is the US enhancing militarism in Africa rather than providing vastly expanded assistance to the beleaguered African Union and other regional peace-keeping and diplomatic organizations? Dressing up AFRICOM in the language of humanitarian and diplomatic engagement fools no one, especially when US military intervention in Africa in the past five years has been so destructive, myopic, and self-interested. AFRICOM will not benefit Africans, it will not enhance African security, and it will not prevent war. Rather, it will benefit the US military, US defense contractors, US oil companies, African governments interested in repressing indigenous, minority, activist, environmentalist, and resistance groups, and terrorists, who will win new adherents to their causes amongst those attacked by US rhetoric and provoked by the presence of US troops on their soil.

If the goal of AFRICOM is humanitarian and diplomatic, as President Bush promised when he announced its creation, the US military is the least appropriate vehicle for offering such support. The spectacularly well funded US military will be able to override the relatively poorly funded efforts of the State Department, USAID, and NGOs in defining US policy in Africa, allowing mission creep and an increasingly militarized response to African conflicts. Given the extremely complicated nature of the current conflicts in the Great Lakes region, the southern Sudan, Somalia, Zimbabwe, and the Niger delta, prioritizing military rather than diplomatic intervention in such places is a cause for grave concern. Because of AFRICOM, it is certain that the localized engagement so fundamental in diplomatic relations and development projects will be subverted by regional plans defined by the military. Already Congress has asked AFRICOM for a comprehensive regional counternarcotics strategy for West Africa. Africans have only to look at Plan Colombia—which consistently prioritizes tools of aggressive military action over diplomatic and humanitarian initiatives—to see what that might look like.

Through AFRICOM's military assistance to African governments, the US will have an even stronger voice in shaping domestic laws and policies within African countries regarding terrorism and resource extraction, as well as greater power to affect the ability of African governments to independently formulate foreign policy and support for the rights of those considered terrorists by the US government, such as Islamic groups, Palestinians, and those whose interests are subverted by US foreign policy elsewhere in the world. In her book *In the Moment of Greatest Calamity*, anthropologist Susan Hirsch provides a distressing portrait of the damaging effects on the Kenya legal system of relentless US pressure to conform the treatment of the men accused of the East African US embassy bombings to US rather than Kenyan law. AFRICOM will offer another vehicle through which such pressure can be applied.

Finally, a US combatant command for Africa has the power to focus attention and resources on military rather than alternative forms of engagement and intervention. Countering the militarization of US foreign policy will be a crucial component of scholarly and activist work in Africa in the upcoming decade.

A Review of AFRICOM Promises

President Bush and Pentagon officials eagerly attempted to cast AFRICOM at its inception as a humanitarian undertaking concerned with the political, economic, and physical security of Africans. President Bush's February 2007 speech announcing the creation of AFRICOM promised, "Africa Command will enhance our efforts to bring peace and security to the people of Africa and promote our common goals of development, health, education, democracy, and economic growth in Africa." Officials described AFRICOM as an "interagency structure" that would include military and civilian staff who would work together on security, diplomatic and humanitarian activities. Senior State Department security advisor Ambassador Robert Loftis told NPR in

February 2007 that AFRICOM represented an effort to "find a way where the Department of Defense, Department of State, USAID, and eventually some of the other players from the federal government can find a way to work together more efficiently." AFRICOM spokespeople promoted the humanitarian projects undertaken by soldiers stationed at the only US military base in Africa, Camp Lemonier in Djibouti, as indicative of the soft power "hearts and minds" counterterrorism orientation of the new AFRICOM. As CENTCOM commander General John Abizaid told *Time* in August 2006, military involvement in

> low level civil projects throughout HOA [Horn of Africa] such as digging wells, building schools and distributing books, and holding medical and veterinary clinics in remote villages... bolster local desires and capabilities to defeat terrorists before they can become entrenched.

Promotional materials and official speeches stress AFRICOM's focus on "development, diplomacy, and prosperity" for security in Africa. The AFRICOM website and press releases, congressional testimony of AFRICOM's leadership, and interviews in the media with AFRICOM officials trumpet AFRICOM's "interagency cooperation," "three pronged defense, diplomatic and economic effort," "partnerships" with NGOs and African nations, an integrated staff consisting of military and civilian personnel, and a "robust," "close," "integrated working relationship with the Department of State." AFRICOM is the first command to have a civilian Deputy to the Commander for Civil-Military Activities in addition to a military Deputy to the

Commander for Military Operations, as well as staff from the Department of State, Commerce, Agriculture, Treasury, and USAID—all, of course, firmly under the control of the Pentagon. In a March 2008 policy speech about how AFRICOM represents an extension of Bush's strong humanitarian commitment to Africa, Deputy Assistant Secretary of State for African Affairs Todd Moss gushed, "History will, I believe, show our policy toward Africa as a shining legacy of the Bush administration."

A Legacy to Reject

It soon became apparent that many of AFRICOM's intended "partners" wanted no part of this legacy and were extremely suspicious of AFRICOM's goals and its purported humanitarian orientation. It is no surprise that AFRICOM's glowing rhetoric of peace-building, diplomacy, and conflict prevention struck Africans as unbelievable, especially since AFRICOM officials offered fuzzy and contradictory explanations of how the Pentagon viewed such things. Although planning for AFRICOM began years ago as part of then Secretary of Defense Donald Rumsfeld's vision for militarizing US foreign policy and covering the globe with US military bases, defense planners never consulted with African nations or with the US Congressional leadership on African affairs. Within weeks of the official announcement about the creation of AFRICOM, members of Congress, NGOs, USAID and State Department officials, academics, and, especially, Africans and African governments expressed alarm and protest. NGOs and academics formed anti-AFRICOM networks[*] and testified against AFRICOM in Congressional committees. Regional organizations of African states pledged to ban US bases from their territories. Every single African country except Liberia refused to host the new command headquarters. The 2009 Fiscal Year House Defense Authorization Bill even expresses concern about the perception that AFRICOM will militarize US foreign policy in Africa.

    Embarrassed by the negative reaction, US defense officials backpedaled away from their initial emphasis on AFRICOM's humanitarian and diplomatic orientation and began promoting AFRICOM's mission as a strategy of "active security" that offered military support for, but not control over, diplomatic, developmental, and humanitarian initiatives in Africa. AFRICOM Commander General Ward announced in February 2008 that AFRICOM would remain based in Stuttgart, Germany (although the US has negotiated "lily pad" access to bases in Uganda, Mali, Senegal, Gabon, Morocco, Tunisia, and Algeria, where a P-3 Orion aerial surveillance station was established, in addition to the base in Djibouti). Ward also emphasized to Congress that AFRICOM will "stay in its lane," undertaking civil activities like building schools and digging wells for visibility and PR benefits rather than taking a leading role in humanitarian initiatives. (In the same testimony Ward asked Congress to ensure adequate funding for AFRICOM's partner agencies because "if interagency capabilities are not better resourced, non-traditional tasks will, out of necessity, default to military elements.") The projected number of civilians working in AFRICOM was dramatically

---

[*] See, for example, http://www.resistafricom.org

reduced from half the total of 1300 to less than 50, and by June 2008 the Commander was describing their role as "giv[ing] a perspective that we might not always have inherent in the headquarters."

AFRICOM officials have been working hard to explain what AFRICOM's focus on "active security" will look like in Africa. Arguing that security is a necessary precondition for development, officials say AFRICOM will be heavily involved in military-to-military training to improve African military capacity. US military officials express concern about Africa's unpatrolled borders and unpoliced areas where terrorists might hide, but insist that AFRICOM's role will be to strengthen African military capacity to police these areas rather than to establish US military bases in Africa to hunt terrorists. As General Ward told Congress in March 2008: "By strengthening our partners through capacity building efforts, we will deny terrorists freedom of action and access to resources, while diminishing the conditions that foster violent extremism."

But many critics are unconvinced. In addition to the particular concerns about AFRICOM's objectives detailed below, AFRICOM's orientation toward military capacity building continues the destructive trend of defining security in military terms. Since the current ratio in US funding for defense versus diplomatic/development operations abroad is 17 to 1, there are grave concerns about the power of the Pentagon to orient US policy in Africa toward military rather than civilian functions. In the African context, strengthening the role of civilians in governance, security, and democratic practice is far more critical than strengthening African military capacity. If US resources are poured into military training rather than civilian capacity building, the unfortunate result will be a militarization of civilian functions like policing, post-conflict humanitarian projects, and development initiatives. Militarizing society works against democratic decision-making, civilian controlled governance, and participatory citizenship.

Counterterrorism in Africa

A major emphasis in AFRICOM's approach to security undoubtedly will be pursuing the Global War on Terror. US defense officials like to say that Africa is the new front line in that war. Claiming that terrorists fleeing Iraq and Afghanistan have made their way into the Horn and across the Sahel and the Maghreb, the US established two significant counterterrorism operations in those regions that will come under the control of AFRICOM. Both operations have orchestrated brutal attacks on civilians, supported unpopular governments, conflated Islamist political groups and terrorism, and enabled rather than reduced the growth of Al Qaeda influence in their arenas of operation. These interventions suggest the extent to which AFRICOM's focus on "active security" will take the form of heavy handed, force-driven incursions in the name of the global war on terror.

The humanitarian activities of the Combined Joint Task Force-Horn of Africa (CJTF-HOA) based in Djibouti, have been widely promoted as an indication of AFRICOM's soft power approach. But the combat activities of CJTF-HOA are far more notable. After the Islamic Courts Union established the first relatively stable governing structure in fractured Somalia since 1990, CJTF-HOA provided military and intelligence support for the December 2006 Ethiopian military invasion of Somalia to overthrow the Islamic Courts Union, which the US suspected might be friendly to Al Qaeda. The Ethiopian invasion utterly destabilized the country, sparking a humanitarian disaster identified by Refugees International as *the worst in the world*. Over 60% of the population of Mogadishu was displaced by the fighting. Following this disaster, the base in Djibouti orchestrated air strikes into Somalia to kill terrorists, which instead killed dozens—and perhaps hundreds—of civilians. US military intervention has contributed to a level of human desperation, insecurity, and refugee flows in Somalia that are worse now than ever before. The military firepower option was chosen against the advice of top diplomats, academics, and foreign policy experts who promoted diplomatic engagement with the Islamic Courts Union rather than war.

Operation Enduring Freedom: Trans-Sahara/Trans-Sahara Counter-Terrorism Partnership (encompassing Morocco, Tunisia, Algeria, Mali, Chad, Niger, Mauritania, and Senegal) provides an equally sobering example of US counterterrorism efforts in Africa. Using the war on terror rhetoric to label local level insurgents as terrorists, the US injected funds, intelligence, military equipment and covert personnel into the area and began using a base in southern Algeria to hunt terrorists in the Sahel. Careful ethnographic research by anthropologist Jeremy Keenan suggests US military involvement in the region fanned the flames of domestic and regional discord and helped create terrorists where there were none before. Human rights organizations argue that the

US global war on terror and military activity in the region helped form "Al Qaeda in the Islamic Maghreb" out of local level domestically-oriented resistance groups. US interest in Algeria is undoubtedly related to its oil resources. Tensions between the Algerian government and citizens over the US military activity in the country and the expenditure of Algeria's energy wealth promise future conflict.

If the activities of these operations are an indication of future US counterterrorism initiatives, Africans have much to be worried about.

In addition to these two counterterrorism initiatives, the US also provides military equipment and training to many African countries. As was the case during the Cold War, such assistance is given on the basis of political and economic interests rather than government transparency, democracy, and human rights records. For example, Ethiopia receives support, even though it used US tanks against its own population; Rwanda receives support even though its intervention in the DRC conflict zone exacerbates violence; Uganda receives support even though northerners continue to suffer human rights abuses by government soldiers. Greatly expanded military support for Algeria and Angola is clearly tied to US interests in energy resources. AFRICOM officials are actively seeking to expand the role of the Defense Department in providing military equipment and training. Because some of these programs were handled by the State Department, a transition to military supervision means such programs will become even less articulated with civilian capacity building and oversight. Disturbingly, some of the military training has been handled by contractors, such as DynCorp International in Liberia, where US contractors rather than the Liberian government determined the size and structure of the Liberian army. There is a serious problem with US security interventions in Africa that provide military equipment and training that subvert civil society's engagement in security issues, support repressive governments, facilitate inter-African conflict, and enable human rights abuses. And there is a really serious problem with US security interests in Africa so clearly tied to energy resources.

Oil/China

Although US officials highlight the goal of countering terrorism in Africa, observers and analysts also emphasize AFRICOM's likely role in securing oil for the US and countering Chinese influence in Africa. AFRICOM officials downplay the former and deny the latter. This reticence to acknowledge what everyone knows about US interests in Africa makes AFRICOM's mission seem particularly suspect to Africans and other observers. The US now imports more oil from Africa than from the Persian Gulf, and Africa will likely provide over a quarter of US imports in the next few years. China's economic engagement with Africa has quadrupled in the past five years, undergirding a new political solidarity with oil rich African nations. Policy analysts at right wing think tanks supportive of AFRICOM hold conferences to discuss China's influence in Africa, the growing importance of Africa as a major source of oil, and the likely imminent global tug-of-war over Africa's resources. Analysts on both sides of the political spectrum suggest that squaring off against

China for African resources is the new Cold War in Africa. Since the first Cold War left such a disastrous legacy of corrupt dictators, insurgencies and odious debt (including some of Africa's most intractable conflicts in the Democratic Republic of the Congo and Somalia, most war ravaged landscapes in Angola and Mozambique, and most heavily indebted poor countries), the promise of a new Cold War is ominous.

Africa clearly fits smack in the middle of the US geopolitical logic that binds together US foreign policy, security, and oil. Africa's increasing importance as a source of oil (to say nothing of Africa's other precious resources such as diamonds, gold, coltan, fish, and lumber) is undoubtedly what provoked Rumsfeld's initial vision of AFRICOM. The same logic led to the creation of CENTCOM in 1980, a dire indication of AFRICOM's likely future. In his book *Blood and Oil*, Michael Klare describes how US interest in oil security in the Gulf led to the provision of military assistance, then the establishment of military bases and the creation of CENTCOM, and eventually to combat. He notes that this trajectory was repeated in the Caspian Sea, and is now beginning in Africa.

Much of the US focus on oil security targets the Gulf of Guinea, Nigeria, Algeria, and Angola. The latter three have received major increases in military assistance in the past few years. The US Navy has greatly expanded its presence in the Gulf of Guinea where it has established a permanent station and carried out several military exercises to practice a speedy intervention in the area. EUCOM commander General James Jones said that US Navy carrier battle ships in the Mediterranean would start spending half their time going down the west coast of Africa. Several top

Defense Department officials have been quoted saying that securing oil resources and oil transport will be a key concern of AFRICOM. The US-based oil companies that have been rapidly expanding their investments throughout West Africa must be delighted.

Official caginess about AFRICOM's goals combined with the sorry history of foreign intervention in Africa suggests nothing good for Africa will come from a militarized focus on Africa by an imperial power rapaciously pursuing resources while touting humanitarian objectives. Africans have heard this all before; the rhetoric of development and humanitarianism justified the armed colonial takeover of the continent for resource extraction a century ago, as did US intervention in African politics during the Cold War. Africans know that US concerns about terrorism, oil, and Chinese influence will define US military engagement on the continent, and the pre-existing US interventions in Africa that now come under AFRICOM indicate what they can expect.

Counter AFRICOM

At the African Studies Association meetings in November 2007, the US ambassador to the African Union, Cindy Courville, delivered a searing address to the large audience of Africanist academics. She claimed that the US government was deeply committed to working toward solutions for African problems and chastised the assembled academics for their failure to offer useful advice and information about Africa to the Bush administration, for their penchant for unhelpful criticism, and for their orientation toward abstract theory. She challenged the academic community to participate in the conversation by offering policy solutions rather than sitting on the sidelines, blaming the history of colonialism and the slave trade for Africa's problems. In response, academics lined up at the microphone to reiterate, one after another, the laundry list of policy suggestions they have vociferously championed for the past decade: debt relief, changes to patent law that would allow Africans to develop medicines at sharply reduced prices, the removal of US agricultural subsidies that kill Africa's ability to compete, reduced weapons sales to African countries, and more. Courville listened in stormy silence.

The interaction made clear the discrepancy between US policy objectives in Africa and the policies that would actually support African initiatives toward greater economic, political, and physical security. To counter AFRICOM in the coming decade, academics should insist on these issues as persistently, loudly, and forcefully as possible.

Scrutinizing how AFRICOM defines and implements security will be a top concern. Academics with detailed knowledge about local conflicts and resource allocation issues should be vocal about promoting non-military interventions and indigenous, multilateral forms of mediation and peacekeeping. If a Pentagon-driven foreign policy emphasizes the global war on terror and oil as primary US security interests in Africa, academics must assert alternative definitions of security that focus on poverty, HIV/AIDS, peacekeeping, human rights, the needs of refugees, and diplomatic conflict resolution strategies. These are profound concerns for Africans that critically impact security on the continent. They are concerns that demand civilian oversight of military activity, democratic capacity building, long-term development planning, and the coordinated involvement of multilateral organizations. Reframing problems away from military solutions toward civilian capacity building, democratic governance, and service delivery (health, education, infrastructure) will be an important academic intervention in policy discussions. Enhancing African military strength in a context of poverty, weak government, poor service delivery, health crises, and low educational levels is a recipe for disaster. Granting militaries the responsibility for post-conflict reconstruction, policing, and development and humanitarian projects weakens civilian structures and democratic life. Furthermore, challenging and countering war on terror rhetoric when analyzing the actions of resistance groups and Islamic political organizations is especially critical.

Countering the recent US tendency toward unilateralism will be another important intervention. If we are stuck with AFRICOM, the greatest contribution AFRICOM could make in Africa would be to support the efforts of the continent's peace-keeping operations. Building military strength of individual states—such as the controversial and phenomenally costly DynCorp International contract to train Liberian soldiers—pales in importance when compared with the challenges facing the beleaguered African Union peacekeepers and UN peacekeeping operations. If AFRICOM's focus is indeed "active security," offering support and coordinated assistance to these multilateral operations should be the number one priority.

Overall, academics opposed to AFRICOM must maintain a consistent focus on the big picture by asking whose goals are being met by AFRICOM and who is benefiting from AFRICOM's activities. Holding a relentlessly critical stance about US foreign policy objectives in Africa and studying how those objectives translate into policies with a military component is important for tallying costs and benefits, winners and losers. Grasping the big picture is also important for confronting academic complicity with US militarism in Africa, whether researchers accept Minerva funding to study Chinese influence in Africa or work with the State Department on capacity building projects with African security forces.

The new spotlight on Africa created by political debates about AFRICOM affords an opportunity to promote realistic possibilities for beneficial US involvement in Africa. In addition to maintaining a critical commentary on AFRICOM, academics must seize this opportunity to deliver a unified message about the importance of debt relief, foreign aid, peace-keeping, expanded refugee admissions to the US, changes to patent laws, reduced agricultural subsidies, and knowledgeable, context-specific diplomacy. Promoting alternatives to militarism will be the most effective strategy to counter AFRICOM.

Part IV
Anthropological Implications

Chapter 7
Anthropology and HUMINT
Andrew Bickford

Anthropology is a precarious endeavor: working intensely and intimately with people around the world, anthropologists are privy to information that illuminates aspects of peoples' lives, but which also has the potential of harming those very same people by exposing thoughts, feelings, structures, laws, rules, taboos, and insights that can be used against them. Unlike disciplines which have traditionally been a part of the national security establishment (such as political science and economics), anthropologists work with living people, and spend a great deal of time with them, often years at a time, and possibly decades over the span of a career. Anthropologists learn an enormous amount about the people they work with; much more than they ever publish. Some goes unpublished because it is potentially sensitive or even dangerous to informants. Because of this, anthropologists follow a code of ethics designed to mitigate harm and hurt to the people with whom they work, people who have given them support, friendship, and perhaps most importantly, trust.

The military is also interested in people, but not in the same way as anthropology, even though it wants to use anthropologists and their methods and insights. *The US Army/Marine Corps Counterinsurgency Field Manual* (published by the University of Chicago Press in 2007) stresses, again and again, the importance of understanding people and sociocultural analysis as key components to a successful counterinsurgency. Military operations are predicated on a knowledge of the enemy, or to use a US military term—"threat" forces: type and composition of forces, weaponry, intentions, morale, logistics, their knowledge of "our" plans, their willingness to stay and fight or cut and run. The military employs a variety of means to acquire this information, analyze it, disseminate it, and draw up plans of actions based upon it. It also employs a wide variety of military and civilian specialists to conduct intelligence operations and analysis. There are many kinds of intelligence operations and collection practiced by the US military: SIGINT (signals intelligence), IMINT (image intelligence), and HUMINT (human intelligence) being perhaps the best known. The closest form of intelligence collection in counterinsurgency operations to anthropology is HUMINT, an intelligence "discipline" focused on exploiting people for information. Because they are closely related, the field manual stresses the importance of HUMINT collection and the analysis of socio-cultural information.

The *Counterinsurgency Field Manual*, and the development of the Human Terrain System (HTS), seek the use of anthropology as a form of HUMINT, and anthropologists as HUMINT collectors. I spent five years on active duty in the US Army, working as a sergeant and linguist in signals intelligence in West Berlin from 1986 to 1989. While I was never involved in HUMINT activities, I did learn about the goals of intelligence collection. Having served in the army, I find the arguments used by the *Counterinsurgency Field Manual* and HTS program about the role of anthropology to be troubling and disingenuous, to say the least. One learns in the Army from day one that your job, and the job of the US military, is to fight, control, dominate, and kill. There are many military specialties at both the enlisted and officer levels that could fulfill the same role as counterinsurgency/HTS anthropologists—Foreign Area Officers, and some of the enlisted intelligence specialties such as 97 Echo (Human Intelligence Collectors) in the Army—to name a few. There are also other governmental specialists, such as State Department Foreign Service Officers, who provide analogous information. I've talked to a number of friends who have been in the military—officer and enlisted, with combat experience—and they've quickly understood that the *Counterinsurgency Field Manual* and HTS are about human intelligence collection in which anthropology is used as a cover. There is not much concern for understanding culture as a non-militarized anthropologist might approach it. The concern is: What can culture do for us, and how can culture make us fight better and more efficiently? As one former Marine officer at the Center for Advanced Operational Cultural Learning is on record as saying: "Cultural understanding is a weapon." Counterinsurgency is just another way of saying occupation, but in a way designed to mask the occupation.

As I considered how to write this chapter, I came up with a series of broadly related questions that I see as relevant to understanding the relationship between HUMINT and anthropology. Why is there a need for a counterinsurgency manual that draws on culture, a group's way of life, its ways of being in the world, and then uses it against them? What is the role of anthropology in the exploitation of the "culture variable?" Or rather, how does the use of culture become a "force multiplier" for a military force engaged in counterinsurgency operations? How does the military's engagement with people—HUMINT—link up with anthropology? Does anthropology contain an inherently dangerous aspect, particularly in relation to HUMINT? What is this, and why must anthropologists be constantly aware of the fine line between ethnography and intelligence collection? Should anthropology be used as a tool that may result in the death and wounding of countless people? Is it right to use anthropology to win a war? Can anthropology really ever win a war? Is it not a sort of red herring, a false Holy Grail of victory, domination, and a "job well done"? And are things really so bad in the military that anthropology is seen as a kind of "wonder weapon"?

The *Counterinsurgency Field Manual* contains a great deal on the necessity and importance of intelligence collection. The importance of intelligence efforts to counterinsurgency operations is stated at the beginning of the chapter entitled "Intelligence in Counterinsurgency": "the ultimate success or failure of the mission depends on the effectiveness of the intelligence effort." Without directly saying it, the *Counterinsurgency Field Manual* highlights the necessity of anthropology:

> Intelligence in COIN [counterinsurgency] is about people. US forces must understand the people of the host nation, the insurgents, and the host-nation (HN) government. Commanders and planners require insight into cultures, perceptions, values, beliefs, interests and decision-making processes of individuals and groups. These requirements are the basis for collection and analytical efforts.

The *Counterinsurgency Field Manual* devotes much space to intelligence collection, specifically to HUMINT collection. Section 3-130 of the manual defines HUMINT as:

> the collection of information by a trained human intelligence collector from people and their associated documents and media sources to identify elements, intentions, composition, strength, dispositions, tactics, equipment, personnel, and capabilities... HUMINT uses human sources as tools and a variety of collection methods, both passive and active, to gather information to satisfy intelligence requirements and cross-cue other intelligence disciplines. Interrogation is just one of the HUMINT tasks... HUMINT operations often collect information that is difficult or sometimes impossible to obtain by other, more technical, means. During COIN operations, much intelligence is based on information gathered from people.

The *Counterinsurgency Field Manual* also lists "Comprehensive Insurgency Analysis Tasks." Among those identified as crucial for counterinsurgency operations are:

- Determine how culture, interests, and history inform insurgent and host-nation decision making.

- Understand links among political, religious, tribal, criminal, and other social networks.

- Determine how social networks, key leaders, and groups interact with insurgent networks.

- Determine the structure and function of insurgent organizations.

- Identify key insurgent activities and leaders.

- Understand popular and insurgent perceptions of the host-nation, insurgency, and counterinsurgents —and how these affect the counterinsurgency.

The *Counterinsurgency Field Manual* also notes that "some of the most important actors in counterinsurgency warfare are not self-identified warriors. In counterinsurgency, civilians and non-kinetic actions become the Soldier's exit strategy."

Anthropologists are envisioned in the ranks of these "most important actors" who will undertake the analysis tasks. We are privy to data that, if used incorrectly, or for purposes other than anthropological analysis, can result in people being tortured, wounded, abducted, or killed. It is specifically this kind of information which is of interest to HUMINT collectors.

Anthropology and HUMINT: Scale, Scope, and Intent

The chapter in the *Counterinsurgency Field Manual* that draws most heavily on anthropological insights and methods is titled "Intelligence in Counterinsurgency." This is telling, as there are troubling similarities between HUMINT and anthropology: both work with people and are interested in gathering and garnering information about people. In many ways, the similarities are very close—uncomfortably close, which is why many anthropologists are troubled by HTS and the use of anthropology for counterinsurgency. Both utilize "informants," gather information, interact with people, observe, collect data, analyze and write up findings and interpretations. To reiterate the old cliché, many anthropologists have their "are you a CIA agent?" story from their fieldwork days.

Because of these similarities, anthropology has a potentially dangerous side, one that we must be constantly aware of: the information we collect in the field can be used against the people with whom we work. Anthropology has been used as a form of control in the past, with unfortunate consequences for both the objects of research and anthropology as a discipline; it is just this use of anthropology that the *Counterinsurgency Field Manual* attempts to resurrect. For this reason, anthropologists must be acutely aware of the ethics of fieldwork and participant-observation: what we collect in the field is analogous to the types of information collected and desired by intelligence agencies, and if they can use what we collect, or co-opt our work for their own uses, they will.

But this is where the similarities end, or rather, diverge. The main difference is one of intent: what does one plan to do with the information collected and analyzed? Is it for peaceful purposes, for advancing knowledge about a group, village, city, civilization, or state? Or is it to be used to further the ends of one state against another? This is a road anthropology has been down before, and should not take again.

HUMINT in counterinsurgency is used at a tactical level, to gain local-level knowledge to understand how, what, and why people are thinking and acting in a given area, and who is either supporting or is part of an insurgency. Of course, this information can pass up the intelligence chain and influence strategic-level thinking, but its immediate intent is local and tactical. HUMINT and HTS operations can have many collectors focused on one group, area, or problem; while academic anthropologists have also attempted this kind of "saturation" in the past, these attempts are miniscule in comparison to the vision of total knowledge and understanding laid out in the *Counterinsurgency Field Manual*, a totality directed towards complete understanding for purposes of control. Whereas anthropologists continue to analyze and rethink their field data, tactical HUMINT is only concerned with what is necessary for the mission or the objective; the rest is superfluous.

HUMINT agents—"collectors," according to the manual—are interested in many of the same sorts of questions anthropologists find interesting; the difference is in intentionality, scale, and scope. There is a vast gulf between the intentions of the military and the intentions of academic anthropology. The military is there to do things, to accomplish missions, to execute and further national policy; it is the "tip of the spear" of national interests and policies, and sees itself as the agent of these policies and interests. HUMINT is directly linked to policy implementation. As conceived in the *Counterinsurgency Field Manual*, anthropology is to become a tactical-level policy tool, a tool to implement the desired goals of the US, rather than a discipline concerned with increasing our knowledge of the world to help prevent war and facilitate the resolution of conflicts.

The HUMINT Ouija Board: Predictive Anthropology and the "Intelligence Preparation of the Battlefield"

Unlike conventional anthropology, which is not a "predictive" discipline, the role of anthropological analysis in counterinsurgency is clearly laid out in Section 3-170 of the *Counterinsurgency Field Manual*:

> Comprehensive insurgency analysis examines interactions among individuals, groups, and beliefs within the operational environment's historic and cultural context. One of the more important products of this analysis is an understanding of how local people think. This knowledge allows predictive analysis of enemy actions. It also contributes to the ability to develop effective IO and civil-military operations.

"Local people" in counterinsurgency operations are viewed with suspicion, as potential "threats" to the occupation whose actions need to be constantly predicted and reliably predictable. Ostensibly concerned with helping and trusting the "locals," in the end, they are seen by the military as a problem to be understood and overcome. The dulcet tones of help, support, and friendship are ultimately belied by operational paranoia and the ever-present question behind all operations: What if? This is perhaps the key problem with counterinsurgency and anthropology: anthropologists do not view the groups they work with as problems, nor their behavior as something that needs predicting or controlling. Indeed, both "sides"—locals and insurgents—are targets of counterinsurgency intelligence HUMINT operations:

> Counterinsurgency (COIN) is an intelligence-driven endeavor. The function of intelligence in COIN is to facilitate understanding of the operational environment, with emphasis on the populace, host nation, and insurgents. Commanders require accurate intelligence about these three areas to best address the issues driving the insurgency.

Lest there be any doubt of why this is important, the manual also states:

> Intelligence preparation of the battlefield (IPB) is designed to support the staff estimate and military decision-making process (section 3-7)... The purpose of planning and IPB before deployment is to develop an understanding of the operational environment. This understanding drives planning and predeployment training. Predeployment intelligence must be as detailed as possible. It should focus on the host nation, its people, and insurgents in the area of operations (AO).

These passages make clear that intelligence collection—including HUMINT—is about preparing the battlefield, making sure that the military has all of the information it needs to fight and win.

The same sections of the manual in which these passages appear (Sections 3-7 and 3-8) also imply that sociocultural and anthropological information are crucial before the start of a war or hostilities; rather than acting as a mitigating factor, these passages make clear that such information is necessary before the start of combat to better prepare for the fight. This is a sobering realization: many anthropologists (including myself) hope that anthropology can be used to prevent war. Of course I would like it if the military actually knew something about the complexity of the people(s) they were about to engage, respected their ways of life and beliefs, and ultimately decided that war is counterproductive. Sadly, however, if we take the *Counterinsurgency Field Manual* at its word, it seems that anthropology is seen as simply another weapon in the arsenal, another "force multiplier" to facilitate the "Intelligence Preparation of the Battlefield."

Anthropology, HUMINT, and "Real Time" Analysis

The use of anthropology—through projects such as the HTS and the melding of anthropology and HUMINT—creates a sort of HUMINT-Light, a more "open" form of intelligence collection and power projection. Rather than having to rely on agents, the military can use the entire community as its agents, "peacefully" collecting information through anthropologists on everyone and everything of interest in the community, in the process spreading the risk to all involved, and the blame and responsibility beyond the military. Perhaps most importantly for the military, it is HUMINT conducted by non-military "collectors," making it—theoretically—more palatable to those targeted, and making it seem that it is, in fact, not intelligence collection at all, but just anthropology.

Unlike the open access to information in academic anthropology, the intelligence analysis chain in the military flows in one direction, and is heavily dependent on "need to know" security guidelines. Of course, we cannot control who reads or uses our work, but we can write a public rebuttal or refutation; this is not possible for military intelligence analysts. Low-level analysts cannot control what happens to the information collected and/or analyzed by them. Once it is in the intelligence consumer chain, it is beyond their control; they cannot later say, "I didn't like what you did with the information I collected, and I want you to change it." Analysts do not control their product. And make no mistake—HTS members are low-level analysts in this sense: they cannot control what happens, or what is done, with their information, reports, and analysis. HUMINT and intelligence collection imply hierarchy, targeting, and operational priorities, a process very different from the autonomy of traditional anthropological fieldwork, where the anthropologist can revise, reinterpret, and correct fieldwork information over time.

The *Counterinsurgency Field Manual* deploys culture to increase warfighting capabilities and aims to remove a hurdle in the prosecution of war and domination, to make it easier. "Cultural knowledge" might in some instances save lives, but it will also take lives, and help in the selection and targeting of particular individuals and groups. Counterinsurgency HUMINT collectors and anthropologists are to produce a form of knowledge—tactical intelligence—in "real time" that can be deployed to further combat operations and control. Anthropology, while conducted in "real time," is more about the careful analysis of its "real time" research after fieldwork has ended, rather than the production of an immediate intelligence product for "real time" usage. Militarized anthropology is thus used principally to enhance combat and intelligence capabilities at the local, tactical, "real time" operational level.

The *Counterinsurgency Field Manual* also aims to offer a form of total cultural awareness, in the process echoing the panoptic conceits of early anthropology and anthropologists: "once the social structure has been thoroughly mapped out, staffs should identify and analyze the culture of the society as a whole and of each major group within the society."

This is very much old-fashioned anthropology: simply map out the entire culture and social structure, and then: "Once they have mapped the social structure and understand the culture, staffs must determine how power is apportioned and used within a society."

Determining who has power is necessary if military authorities are to utilize, co-opt, or undermine those in power. HUMINT collectors are also to realize the importance of establishing a network of informants from whom they will (hopefully) collect a steady stream of important and useful information on their community, according to the *Counterinsurgency Field Manual*:

> Establishing a reliable source network is an effective collection method. Military source operations provide the COIN equivalent of the reconnaissance and surveillance conducted by scouts in conventional operations. HUMINT sources serve as the "eyes and ears" on the street and provide an early warning system for tracking insurgent activity.

Establishing networks—and figuring out social networks—are part of what anthropologists do. In the conception of the *Counterinsurgency Field Manual*, anthropologists are to serve as unconventional warfare operators, as the early warning system and scouts of the occupation. The HTS, as a non-military HUMINT team, can create these networks using anthropological research methods, and lend a non-military, "soft power" appearance to HUMINT collection.

The *Counterinsurgency Field Manual* does attempt to complicate the situation, but in the process, reveals the scope of the collection effort, making everyone in the "host nation" a target:

> What makes intelligence analysis for COIN so distinct and so challenging is the amount of socio-cultural information that must be gathered and understood. However, truly grasping the operational environment requires commanders and staffs to devote at least as much effort to understanding the people they support as they do to understand the enemy. All this information is essential to get at the root causes of the insurgency and to determine the best ways to combat it.

Not only are the insurgents the target of intelligence collection, but the "people they support" are targets as well, as they are potentially part of the "root cause" of the insurgency. Rather than blaming the people they claim to support, perhaps the root cause of the insurgency should be seen as the occupation.

Countering the Manual: Anthropology, Intelligence, and Ethics

What is troubling about the counterinsurgency manual, and the HTS teams in particular, is the attempt to make anthropology an intelligence-producing discipline, to forcefully bring anthropology into the national security fold, to co-opt professional anthropologists, and to arrogantly dismiss the concerns of professional anthropologists as naïve, or worse. Because we work closely with the people we study, anthropology does not lend itself to national security deployment in the same way as other disciplines do, especially those removed from interaction with living people. The hubris and arrogance of the *Counterinsurgency Field Manual* are also troubling: the tone of certainty of success, of anthropology-as-panacea for counterinsurgency woes. This is troubling both for anthropology, the people of the "host-nation," and US soldiers caught between a rock and a hard place as they try to squelch an insurgency, killing, wounding, getting killed and wounded in the process. The assumption is that cultural knowledge and the use of anthropology can tilt the scales towards victory and help sharpen the "tip of the spear." It is this element of harm that poses particularly serious ethical problems. Rather than providing knowledge that will result in "victory," the boilerplate promises, anodyne assurances, and old-fashioned anthropology of the *Counterinsurgency Field Manual* will only prolong the fighting, resulting in more death, mourning, and despair.

Some say that it is ethical for anthropologists to assist in counterinsurgency operations in order to bring wars to a successful conclusion. However, if anthropologists will be expected to assist the military and the Department of Defense in all operations and in all endeavors, this would be a de facto militarization of anthropology. The history of other disciplines, as well as our own experience in World War II, shows that it is difficult to leave the national security fold once wrapped up in it.

Perhaps the greatest contribution anthropology can make in response to the *Counterinsurgency Field Manual* is to provide critique, i.e. to ask difficult questions, and to question the unethical use of research findings. We need to resist the use of anthropology as a tactical tool, and in the process prompt those in power to ask the difficult questions that their mandate often precludes them from asking, or precludes them from considering in the first place. There are times when anthropology does not have to be oppositional, but in this case, when facing a hostile merger, I think we need to push back and intensively question what many of us see as unethical uses of our discipline. We would all like anthropology to be "relevant," but assisting in counterinsurgency operations is not the path to relevance. A "relevant" anthropology would contribute to the prevention of war and domination, not serve as a tool and enabler of war and domination.

I can partially understand the motivations of the comparatively few anthropologists engaged in counterinsurgency operations, or who join the HTS teams. Howerver, there is a big difference between concern for soldiers and supporting an occupation, and often, the help proffered to soldiers is not seen by them as any help at all. The HTS may in fact be more of a problem for units to which they are attached, as it means more untrained civilians to protect, and another thing to worry about during combat operations. I have no desire to see US soldiers killed or wounded, or deployed in unwinnable and unethical situations. I also have no desire to see civilians killed or wounded by these very same soldiers. If we think of the US military as a labor market, and its soldiers as workers, these are people who find themselves in coercive and exploitative situations. The co-opting of anthropology by the military is just another way the military can compel soldiers to fight and soldier on; it is a form of labor rationalization, a sort of cultural Taylorism that ultimately does very little for the soldier.

As much as the military stresses its code of honor and ethical stance, I find it troubling that the military—and unfortunately, the few anthropologists who decide to work for the military in a HUMINT capacity—do not understand or wish to subscribe to our ethical codes, of doing no harm to the people with whom we work, and of not using the information we gather for coercive ends. The military prides itself on its adherence to a moral and ethical code of conduct. It's time the military respects the codes and ethics of anthropology, and realizes that others hold their ethics and discipline just as dear, and will not go gentle into that good night.

Chapter 8
About Face! An Anthropology Student's Reflections on Militarization
Kanhong Lin

"The purpose of life is to give back to society," said my mother during one of our arguments. "You must work, you must feed yourself, you must find a job. You have to support yourself before you can help others."

For my family, the theme of work dominated our existence as we made the transition from being farmers and laborers in Taiwan to an immigrant family in Silicon Valley. In the rare times we congregated together for family banquets, politics always lit up dinner with stories about life under Japanese occupation in Taiwan and the subsequent martial law under the Kuomintang. Fleeing these troubles, we curiously found ourselves living in a country actively occupying other countries. It was now 2003 and I was ready to drop out of the university, plagued by a restless yearning to do more with my energies than to just work. Driven by this desire, I almost joined the military with the hope that it would allow me to help people.

I started university as an electrical engineering student, a pragmatic choice influenced by my family's desire to see me succeed in life in a "stable" career in law, medicine, or engineering. I was resistant to the idea of becoming an engineer when my real interests lay in history and social science. However, I didn't know what else to do so I sleepwalked through my classes as a passive form of resistance. I was the eldest child and burdened by a sense of filial duty to comply with my family's wishes. Whenever something needed to be done, I was expected to drop everything and get it done. I clashed with my family and relatives over my direction in life. They expected me to study hard, get a job and live a quiet life free from political unrest. Yet I resisted because an unknown desire gnawed at me; I was seeking a greater purpose.

I made up my mind to go to the Army recruiting office to sign enlistment papers. My father was proud of his service in the Taiwanese Air Force and part of me felt it would be a good idea to follow and enlist in the US armed services. As I grew weary of daily arguments and ran out of ideas, the military sounded like a good alternative to inaction. Meanwhile, the Bush administration had been making the case for a war against Iraq to the American public and the world. Fortunately, my cousin convinced me to wait until the end of the semester before joining the military. I heeded her advice and in a stroke of good luck, dropped by the office of an anthropology professor. Our resulting conversation inspired me to stay in university, and soon I was asking fundamental questions about the world and my place in it.

"Cultural Knowledge" and the Military

One of my questions involved reassessing the relationship between society and the military. After all, I grew up with the media glorifying soldiers in movies like *Saving Private Ryan* and video games such as *America's Army*. My generation saw the return of soldiers as cultural heroes and saviors, a trend creeping into anthropology too. The Human Terrain System project (HTS) appears to represent a shift in attitude on the part of the military. On the surface, applying cultural knowledge to guide the military's efforts sounds like a very pragmatic solution—a way of performing real acts to help real people. My decision to join the military stemmed mainly from an innate desire to do something constructive, to give back to society. Feeling as if I could tap some sort of potential, I thought that the military would allow me to help people. I did not want to spread violence but to reduce it. Over the past few months, I have been struck by similar sentiments expressed in public statements by HTS members:

    1. In a written statement to the Project on Government Oversight (POGO), religious studies graduate student Zenia Helbig, a former HTS employee (and now critic) wrote,

> I came to government work as a result of the September 11th attacks. As an immigrant, my father had a deep sense of national service which he passed on

to me, and I felt compelled to do my part. With a Bachelor's Degree in comparative Religion, I believed that the best way for me to contribute was to return to school, gain an advanced understanding of Islam and apply that knowledge to helping the government face this challenging new environment.

2. Writing to *The Chronicle of Higher Education Review*, anthropology professor and Human Terrain Team (HTT) member Marcus B. Griffin wrote,

> Whether you think the United States should have entered Iraq by force (which I don't) and toppled Saddam Hussein, the inescapable fact is that we are here. Now academics have a choice: We can apply our specialized skills in the field to ameliorate the horrors of war, stem the loss of both American and Iraqi lives, and improve living conditions for Iraqis, or we can complain from the comfort and safety of the faculty lounge.

3. Mark Dawson, an applied anthropologist and HTT member, recently blogged about his motivations to join,

> For me, I feel like after all my time in the corporate world, I have something to give back, I hope... Surely I can turn that skill to something more meaningful, and a longer lasting effect? ...Do I believe in the power of cultural understanding to prevent violence or not? I do. Of course, I might be wrong. That's just the way it goes with human endeavors.

4. In a *San Francisco Chronicle Magazine* profile of Montgomery McFate, a cultural anthropologist who is among the key architects of HTS, she stated, "I

wanted to do something in the world, not about the world."

The themes of helping people, decreasing casualties and improving living conditions for Iraqis all run through these statements. Those quoted say they joined up in an effort to reduce civilian casualties, as well as casualties among US troops in the war. These rationales appear extremely seductive at first glance and form the main thrust of the HTS recruitment effort. With anthropology in hand, the recruiters argue, US soldiers might safely carry out missions across the world. In the *San Francisco Chronicle Magazine* article, McFate, in declaring inaction unethical, calls out to anthropologists to directly engage with the military and policy world. Her call is tied to a sense of justice shared by most anthropologists. However any discussion regarding imperialism remains absent. After all "military doctrine is not meant to provoke philosophical foxhole debates" about the war, says McFate. David Kilcullen, another social scientist involved in counterinsurgency, echoes this sentiment by stating, "elected political leaders decide the justness of a war at its outset: properly so, since this is a question for the whole nation, not for military professionals (or any other kind of specialist)." Substitute specialist for anthropologist. Apparently there is no room for dissent when working for the military; instead, argues Kilcullen, anthropologists should focus on executing the war in an ethical manner, but given HTS's refusal to address any of the complex ethical issues raised by its practice, this admonition rings hollow.

McFate's moral call to action works extremely well because it erases the political and historical contexts of the war. As someone once remarked to me after reading a *New York Times* article highlighting

HTS efforts in Afghanistan, "What's wrong with helping widows?" I found this question unsettling as it was a difficult argument to counter. As I talked to people, the idea of saving widows in foreign places superseded any debate about unilateral foreign policy. It got me thinking, "What was wrong with using anthropology to improve the lives of others?" I realized that by appealing to our moral desire for justice, such arguments convince us to set aside our political objections to the military and the war. Our guilt can be assuaged by the notion that saving widows or helping out needy villagers justifies our presence, and that our work is making a difference in the lives of innocent civilians. In reality, our moral outrage and longing for social justice has been co-opted and used as a recruitment tool against us.

An Ethnocentric Anthropology?

In 2006, I moved to Washington, DC to begin working towards my PhD in anthropology. I was excited to be in the nation's capital, a hub of power and policy. For years, many anthropologists have argued for applied anthropology as the solution for a growing set of problems. It was not long before I met a variety of anthropologists working in various DC establishments. Introductions were often prefaced by the exchange of business cards, a ubiquitous form of "gift exchange" in the Beltway.

After all the years of anthropologists grumbling about their insignificance, I had stumbled across a "lost tribe" of anthropologists already working for the government, NGOs, and private sector organizations. If their goal was to provide expertise to craft better policy, it apparently had not worked yet. Instead of anthropologists having a tremendous affect on public officials, it was much more the other way around. The institutional force of Beltway politics ensures that anthropologists often must comply with the rules of the game or be quickly marginalized.

I once found myself in a meeting full of applied anthropologists that later devolved into a complaint session about being ignored and unable to get people to listen to their recommendations. I made a suggestion. Perhaps instead of butting our heads against the bureaucratic wall, we might try other means to communicate anthropologically informed views—such as in the public forum. After all, there is more to society than policy makers, anthropologists, and target communities. Resounding nods and murmurs of approval filled the room—except one problem. Someone asked, "How do we do it? Is there a guide out there on how to write op-eds?"

Some applied anthropologists seem uncritical of their ethnocentrism and often times appear more preoccupied with selling anthropology as a tool for "emerging markets," whether it is in the public or private sector. They often chase after the latest grants, fellowships, sponsorships, or job offers. After all there is a great deal of money to be had. Recently an anthropology student enthusiastically told me about the latest funding for anthropologists offered by the Pentagon—the Minerva Consortium. What has evolved now is a form of imitation anthropology: it smells, sounds and tastes like the real thing. "Anthropology-lite," the late William Roseberry once called it.

I am not condemning all forms of applied work. In fact, many anthropologists—academic, applied and military—have adopted an uncritical view of anthropology's role in the world. It is a view laced with Western exceptionalism, a belief that the West can provide the solution to problems in faraway lands. It is a patronizing view that holds people in non-Western societies as infantile and in continual need of developmental aid or military intervention. But people are not helpless; they understand their condition and actively take control of their own affairs within the constraints of their environments. However, many of us get caught up in playing a parental role whether through the mechanism of the military, top-down development agencies, multinational corporations, or even academic analyses.

Aside from the ethical issues raised regarding embedded anthropological work for the military, there is an important point in all of this. The fact is that no one likes to be colonized or subjected to military occupation, no one likes to feel violated and restricted, no one likes feeling like a powerless pawn. Would we want the same to be done to us? As strange as this idea may seem in policy circles where American power must be preserved at all costs, the truth is that our political, economic, military and social actions often hurt other people.

In an effort to try to gain the ear of policymakers, too many of us anthropologists tout our knowledge, expertise, and research methods as solutions to political blunders. However, anthropology is more than simply research methods. We are nothing without the relationships we cultivate with research communities. We have forgotten that our most powerful contributions have been made when we angle our gaze back upon our own society. Anthropology should not be obsessed with the exotic, esoteric, or the pragmatic but should be informed by cross-cultural understanding of different ways of living. Comparison exposes the contradictions inherent in all societies, a particularly useful way for those living in relatively powerful societies to question assumptions that are taken for granted. Many of us—not just anthropology students, but practitioners and academics—have forgotten the legacy of anthropologists such as Margaret Mead, Ashley Montagu, Franz Boas, Gene Weltfish, Kathleen Gough, and Eric Wolf, all of whom developed powerful critiques of our society. This amnesia weakens the practice of anthropology and future generations of anthropologists.

War for a New Generation

Fast forward a year. After discovering that I was the only student in my cohort without a graduate stipend, I spoke up about my situation and soon found myself standing in the hallway one day peering at one of my professors. He had stopped me to discuss my funding opportunities.

"Kanhong," he started. "I want you to know that I am pursuing all options to get you some funding."

He paused. "In case I am unable perhaps you should consider applying for a Boren, if you are amenable to the idea. We've had other students awarded the fellowship in the past."

I thought to myself, "Is he joking or is he testing me? Surely he knows my position regarding working for security agencies." The professor knew about my work pushing for the passage of resolutions condemning the Iraq War and torture through the American Anthropological Association meetings. His suggestion seemed disingenuous until I realized he was serious. I had researched the fellowship earlier when another professor in the department made the same suggestion. I found that the David L. Boren Graduate Fellowship was administered by the National Security Education Program (NSEP), a body providing scholarships and fellowships to university students who conduct research on topics critical to US national security. Funding requirements stipulate that after graduation, fellowship recipients must seek employment with agencies affiliated with national security—the CIA, FBI or other similar agencies.

"You might have some problems with the service requirements but some students have been able to get out of the requirement," he continued. I felt conflicted and helpless. I understood clearly that there would not be any further department support. However, I was unwilling to align my research interests with the fellowship's national security theme. More importantly, I felt uncomfortable knowingly applying for a grant with the intent to renege on the contract through some sort of loophole. Breaking the contract would have legally required me to repay the fellowship amount if I did not fulfill my service. My concerns were not unfounded as David Price recently wrote about Nicolas Flattes, an anthropology student, who has been pressured by NSEP to either pay back his loans immediately or fulfill his service requirement. Faced with no prospect of increased department funding, I decided to leave the program rather than pursue funding from the national security state or take out a loan and face an uncertain future.

In a 2006 Survey of Earned Doctorates conducted for the National Science Foundation and other federal agencies, the summary report found that

> three-fourths (74 percent) of the 2006 doctorate recipients reported the primary source of support during graduate school as program- or institution-administered sources, such as teaching assistantships, research assistantships/traineeships, and fellowships/dissertation grants.

Although a large proportion of graduate students are reliant on support from graduate programs, the support has not always been forthcoming. Graduate programs are increasingly accepting more graduates, a strategy used by departments as they push for resource allocations from university administrators. However, these students are left without any funding and must figure out how to meet the costs of higher education. It is unsurprising that some turn to security funding.

With over 500 PhDs (and even more master's degrees) in anthropology being pumped out each year, graduate programs that emphasize quantity over quality of training are endangering the discipline. Master's students have often decried being treated as nothing more than "cash cows" for departments, being pushed through without any training or guidance and left wondering the value of their degree. Sometimes doctoral students fare no better. The "sausage factory" has left anthropology with a large pool of graduates applying anthropology in careers that tread uncharted ethical terrain. William Roseberry's critique of the political economy of American anthropology departments (fittingly entitled, "The Unbearable Lightness of Anthropology") should be updated and extended to graduate students and the production of knowledge. We need to ask ourselves hard questions such as:

- Should faculty push students to apply for security scholarships?

- Should there be ethical guidelines for graduate research funding sources?

- How do security-centric scholarships affect dissertation topics?

- How does social class in graduate cohorts play a role in the production of knowledge?

- How should graduate programs be re-structured to not only train students to be academics but also to effectively engage with pressing public issues?

- How does our over-production of graduate students affect the practice of anthropology?

These trends have a greater effect on the academy than is obvious. Security funding not only potentially biases research but it affects generations of future academics.

Breaking the Cycle

In an article entitled, "The Phantom Factor" (a reflection on the Cold War's influence on anthropology), Laura Nader quotes Jack Stauder in his analysis that anthropologists will not hesitate to serve the US government, corporations, or foundations if their mission is seen as beneficial to things such as social change, democracy or freedom. Amid all the talk regarding global warming, war, terrorism, nuclear threats, oil shocks, and economic collapse, it is understandable that anthropologists want to use their skills to try to ameliorate these problems. Instead of feeling helpless about the situation, we want to make the world a better place. But whose world are we talking about? The world of people struggling to cope with the effects of our government's ethnocentric foreign policies or our own personal world that allows us to feel better about ourselves? There is a danger in uncritical anthropology, be it for the military, for seemingly innocuous civil projects, or for academic audiences.

The past few years have pushed the general public to adopt a dismal view of the government. But rather than seeing any serious opposition mounted against perceived injustice akin to the social movements of the 1960s and 1970s, many Americans have resigned themselves to passive apathy. Are we jaded and cynical about the possibility for change, tired of waiting for responsive leaders, or simply out of ideas? I am more inclined to think we need a new perspective to inspire us with new ideas for change. Contrary to arguments made by people like Montgomery McFate, most anthropologists are not simply sitting smugly in ivory towers sniping against any sort of public engagement. There are anthropologists actively researching and writing about important topics facing our society. They include Laura Nader, William Beeman, Lesley Gill, and David Graeber, along with the contributors to this volume and many more. There remains much work to be done, but a small thriving movement already exists.

What direction should this movement take? Anthropologists can participate in shaping public policy in direct and indirect ways. By utilizing individual strengths, anthropologists can create change through all venues, whether it be writing and speaking to the public, making films, consulting, community organizing, academic research on public issues or education. But the bigger question is whether anthropologists can be involved in national policy without creating more problems. I think the answer is yes—but only if we push for an "anthropolicy," broadly defined as a political policy centered on people, communities, and their needs. Anthropologists can potentially serve as facilitators trying to build dialogue and consensus between communities and politicians. Whatever form of public participation that anthropologists choose, it must be focused on building relationships.

Here are some ideas to help anthropology move towards building relationships with the public:

1. Anthropology needs to assume a greater role in public discourse. Op-Eds and media interviews only reach a certain segment of the public. The American Anthropological Association's (AAA) newsletter deals with anthropology from the perspective of anthropologists. There is a pressing need for appealing, publicly oriented periodicals and books that

deal with current issues in a thorough and critical fashion.

2. Similarly, the AAA should organize public speaking engagements in major metropolitan and university centers in the country. Museums, organizations, or community groups are often eager to schedule talks. These engagements are very good opportunities for anthropologists to share research and analyses while make connections with community members. Better yet, do it yourself. Why wait for an institution to solve a simple problem?

3. Graduate ethics courses need to be introduced or revamped in an effort to involve graduate students in current disciplinary debates. Often times graduate students progress in a vacuum regarding important developments within the discipline. Student voices in the debate are critical in determining future directions in anthropology.

4. David Graeber's treatise arguing for an anarchist anthropology provides one vision for an anthropology engaged in very pertinent issues in our world. The past seven years have forced many Americans to question undivided trust in dominant institutions in our lives. This provides an opening for anthropological work that questions our given assumptions about society and the state. In *Fragments of an Anarchist Anthropology*, Graeber notes that,

> Anthropologists are after all the only group of scholars who know anything about actually-existing stateless societies; many have actually lived in corners of

the world where states have ceased to function or at least temporarily pulled up stakes and left, and people are managing their own affairs autonomously.

5. Introduce humility, reduce ethnocentrism and increase collaboration with other disciplines. I constantly met graduate students who made snide remarks about other academic disciplines: "Anthropology does it best." If graduate students and anthropologists are ignorant of their own history, diverse research methods and a firm grasp of foreign language skills, what exactly do we do best?

6. Pare down the size of the annual meetings of the AAA. Conferences should be sites for debate and discussion about pertinent issues—not dog and pony shows. Open up and invite the public to participate in the meetings. Revitalize regional associations as alternative places to present research. It provides an intimate forum to spend time critically evaluating topics, in addition to the environmental benefit of reducing the huge carbon footprint associated with academic conferences.

Like so many anthropologists before me, I believe in the potential of anthropology to enact change. I believe in its power to wrest a mind free from narrow thinking and its power to make connections between disparate communities. What excites me as an anthropologist is the desire to share my knowledge with people, breaking down walls of ignorance and sharing together a new world full of possibilities. I am saddened that in a moment when anthropology's strengths are most needed, many anthropologists have decided to join the military largely to help accomplish the misguided goals of policymakers. Instead of seeking alternative ways to engage with the public, some have chosen to remain silent, ignore the public, or talk only to policymakers. In the continuing turmoil people have asked, "Where are the teach-ins? Where are the protests?" Now more than ever, I believe anthropologists need to step in to fill the vacuum of ideas. Some anthropologists have decided to step forward to speak and interact with the public. Let us now make strides towards real acts helping real people.

Part V
Alternatives

Chapter 9
Proposals for a Humanpolitik: Building a New Human-Centered Foreign Policy
David Vine

In perhaps the *Counterinsurgency Field Manual*'s most revealing passage, Sarah Sewall, who wrote the introduction to the edition published by Chicago University Press, admits that the manual's ideas are "anything but new." The manual "embraces a traditional... British method of fighting insurgency," she explains. "It is based on principles learned during Britain's early period of imperial policing and relearned during responses to twentieth-century independence struggles in Malaya and Kenya," as well as lessons learned from the dying days of the French empire.

Such an admission should give people in the United States in the 21st century pause. Why, we may ask, is our country modeling its foreign and military policy on the tactics of 19th and 20th century empires bent on dominating colonized peoples? Why, for that matter, is the United States, like Britain and France before it, occupying other nations and facing violent insurgencies in the first place?

An honest discussion of these questions acknowledges that contemporary US counterinsurgency strategy has likewise been built on the US's own history of imperialism and counterinsurgency, from the suppression of anti-colonial insurrection in the Philippines at the start of the 20th century to the war in Vietnam to the wars in Central America in the 1980s. After all, the United States has been an empire for most of its history: conquering lands and peoples in North America and islands outside the continent in the nineteenth century, and since World War II depending less on controlling foreign lands and more on exercising economic, political, and military forms of power, including a massive collection of around 1,000 military bases outside the 50 states and Washington, D.C.

Despite its pacific-sounding talk about "non-kinetic operations" and "armed social work," counterinsurgency is, as Sewall's introduction acknowledges, a form of violent military occupation, not a new and improved vision for US foreign policy.

Counterinsurgency's proponents suggest that applying the lessons of past empires will allow the US to correct the mistakes of its wars in Iraq and Afghanistan. But the lesson of these wars—in nations where the British failed to impose colonial rule and where tens if not hundreds of thousands of Iraqis, Afghans, and Americans have already lost their lives—is not that we need to embrace counterinsurgency and fight smarter, is not that we need to know more about the cultures of our enemies, is not that we need to embed anthropologists with our troops. The lesson is that wars like these, wars where the US is invading and occupying other nations, should not be fought at all.

Our nation is at a crossroads. We can continue a foreign policy of unilateral "preventative war," of a "war on terrorism" that has created more who would wish us harm and made the country less safe, of torture and "extraordinary rendition," of wanton disregard for international law, beginning with the Geneva conventions, of the destruction of domestic civil liberties and *habeas corpus*, of indefinite detention, of out-of-control military spending exceeding $1 trillion a year that's bankrupting our economy.

Or we can reject this foreign policy of invasion, occupation, and empire, and embrace a fundamentally different foreign policy, a different way of engaging with the world.

I believe we must choose the latter course. Using this moment of crisis to transform our foreign policy, we must reject shortsighted and fraudulent notions of "national security" and foreign policy "realism" and instead embrace a new foreign policy of "humanpolitik"—a human-centered foreign policy based around diplomacy, international cooperation, non-aggression, and the protection of human security as the best way to protect the security of the US and, ultimately, the world.

In what follows I outline the beginnings of this humanpolitik and nine main platforms for a foreign policy that provides clear alternatives to counterinsurgency.

Rejecting Realism, Embracing Humanpolitik

A humanpolitik begins by placing human beings and the protection of human lives at the center of our foreign policy. It continually asks, What are the human consequences of the nation's actions not just for US citizens but also, given the increasing interconnectedness of the world, for citizens of the globe? Rooted in the anthropological challenge to see the world from the perspectives of others, this fundamentally different mode of engagement with the world focuses on securing human needs as the best way to ensure the security of the nation and the world.

At the heart of a humanpolitik are the principles of non-aggression and international cooperation. This means resolving conflicts through diplomacy and negotiation rather than invasion and occupation. Non-aggression also means ending covert CIA operations, like those that helped to install the Shah in Iran in 1954 and overthrow a democratically elected government in Chile in 1973. All too often these secret activities have caused tragic unintended consequences, known as "blowback," that have ended up harming the United States and global security—the most obvious case being the CIA's funding of Osama bin Laden and other militant radicals to fight the Soviet Union in Afghanistan in the 1970s and 1980s. With a more focused and less overstretched military that would still provide the most powerful *defensive* force in the world, the nation would reserve the right to self-defense, enshrined in the United Nations charter, but only invoke it in the event of an immediate threat to the 50 states. Given that there will be limits to the effectiveness of diplomacy and international cooperation, the US would also remain prepared to participate in peacekeeping operations and military operations undertaken as part of UN-authorized multinational forces.

But what about protecting ourselves from terrorism? some will appropriately ask. Isn't counterinsurgency the best way to defeat terrorism and win the "global war on terror"? Analysts now increasingly agree that there can be no military solution to the terrorist threat, and that terrorism should instead be treated as a crime best confronted by law enforcement, not the military. Echoing the work of anthropologists Jane and Peter Schneider, among others, a 2008 report by RAND Corporation terrorism experts concluded that the United States should pursue a counterterrorism strategy emphasizing "policing and intelligence gathering rather than a 'war on terrorism' approach that relies heavily on military force."

Embracing a humanpolitik next means changing how we think about *security*. For years, empty invocations of "national security" have been a hallmark of a "realist" foreign policy that has most often ensured the security of corporate and bureaucratic elites. A humanpolitik incorporates a broader notion of security captured by the emerging concept of human security, which focuses on ensuring the ability of all humans to achieve fundamental freedoms from want and from fear. "A modern concept of national security demands more than an ability to protect and defend the United States," Gayle Smith writes for the Center for American Progress. "It requires that we expand our goal... to focus not only on the security of nation states, but also of people, on human security." In a world where World Bank and other statistics show that nearly half of today's six billion people live on less than two dollars per day, billions lack basic human security, threatening the species and making the world a dangerous place. As many (including members of the military) are predicting, global climate change, environmental degradation, disease, famine, rising refugee flows, growing inequality, and the exhaustion of critical resources will make the world more dangerous in the decades to come, paving the way for future wars. Left unattended, "human insecurity feeds on itself," Smith says, "laying the ground for conflict and the extreme vulnerability that causes people to fall over the economic edge." In the face of these challenges, a humanpolitik insists that the security of the US and its citizens depends on dramatic new investments in human security at home and abroad.

This won't be easy. Aggressive military solutions are almost always the easiest and most politically appealing. Embracing a humanpolitik approach will, paraphrasing David Halberstam's analysis of the failure of the "best and brightest" generation's war in Vietnam, require much political risk and a deeper, more interior form of strength and courage than that demonstrated by using military force.

In part for this reason, a humanpolitik rejects the traditional practice of foreign policymaking in which an elite group of government officials and think tankers craft policy based on their own narrow (often big business-shaped, economically-driven) perceptions of what's realistic. In its place, a humanpolitik emphasizes the democratization of policymaking, increasing grassroots involvement in foreign policy, and the consensual determination of the nation's interests. Although the idea that there is a singular "human interest" is potentially as flawed as that of national interest, focusing on human needs can help to democratize foreign policy, bringing it closer to representing the interests of the many rather than just the elite few.

With two deadly and hugely costly wars dragging on, with the US economy in a recession and dangerously close to collapse, with the nation's debt reaching $11 trillion, the writing is on the wall for the US as an empire. We can choose the same path the British and French empires took in the 20th century, squandering more lives and dollars to maintain our empire, further bankrupting our nation and setting ourselves up for ignominious defeats like those at Dien Bien Phu in 1954 and Suez in 1956, or we can avoid such tragedies by giving up our empire and embracing a model of global cooperation.

Some will still insist that a humanpolitik approach isn't realistic. The dead in Iraq, in Vietnam, and the dead across more than half a century of an interventionist foreign policy that has brought destruction abroad and helped create unparalleled bankruptcy at home demand that we ask, "Realistic for whom?"

Building a Humanpolitik

To further define a vision for a human-centered foreign policy I describe nine main platforms for a humanpolitik today. Each of the platforms is necessarily schematic but represents an outline for a different and eminently realistic way of engaging with the world. Each draws on the work and ideas of others, many of whom have far more detailed proposals than space allows here.

There are certainly important foreign policy positions omitted here that should be included in a broader elaboration of humanpolitik.

Iraq and Afghanistan. Shifting the nation's foreign policy must begin with repairing the gravest error of recent years, Iraq. First, all US and other foreign troops should withdraw from Iraq with all due speed as requested by the Government of Iraq and consistent with Iraqi and US public opinion polls. The withdrawal must include the closure of all US and other foreign military bases in the country. Second, the US should commit to building cooperative diplomatic and political solutions to maintain peace. As many others have suggested, a withdrawal must be accompanied by a major diplomatic initiative, under the auspices of the United Nations and perhaps the Arab League, to initiate a ceasefire among all combatants in Iraq and to advance political and economic power-sharing arrangements in the country. A parallel regional diplomatic initiative under United Nations and Arab League leadership should seek to create a regional non-aggression pact to support the future stability of Iraq and the greater Middle East. The US must further commit to being the lead funder for an international reconstruction effort for Iraq. This Iraqi Marshall Plan should be directed by the Iraqi government, not the US or US corporations now engaged in reconstruction, and provide the government with significant reconstruction funds over at least a decade as a form of war reparations. The plan should focus on employing Iraqi rather than US or other foreign citizens and contractors and should provide assistance to displaced Iraqis within and outside the country.

In Afghanistan, the growing strength of the Taliban, rising civilian deaths from US military actions, and the failure of the British and Soviet empires before us signal that there will be no military solution there either. As repugnant as the Taliban is, engaging its leaders in political dialogue as part of international diplomatic and reconstruction initiatives similar to those in Iraq will be the only way to build peace and human security for the Afghan people. At the same time, US and international law enforcement agencies should continue to pursue the capture and prosecution of Al Qaeda members and other terrorist criminals in Afghanistan and Pakistan as part of the kind of police-based counterterrorism strategy the US should have pursued immediately after September 11, 2001.

Cutting Wasteful Military Spending. US military budgets, driven by the military-industrial complex President Eisenhower warned us against, have grown larger than they were during the cold war despite the absence of a threat on the scale of the Soviet Union. The Pentagon's official budget for the 2009 fiscal year is $515.4 billion. Adding budget requests for Iraq and Afghanistan and military and defense-related spending in the departments of Energy and Homeland Security and other agencies, the real US military budget is approximately $1.2 trillion. Such spending is obscene. It endangers our security by threatening the bankruptcy of the nation and the collapse of our economy.

The United States must substantially reduce the size of its military budget and refocus on defending the actual territory of the country. The US has the most powerful military on earth and with this proposed contraction can still comfortably maintain this position.

Researchers at the Institute for Policy Studies and the National Priorities Project identified cuts to the Pentagon's 2008 budget that would result in a savings of around one-third, or $213 billion. The cuts would come from an Iraq withdrawal, the elimination of wasteful and redundant weapons systems (often pushed by Congress and unwanted by the military), and a one-third reduction in military personnel and the approximately 1,000 US military bases outside the 50 states and Washington DC, among other savings. Under the plan, the US would still have a military budget more than eight times larger than the world's next largest militaries in Britain, France, Japan, and China. Most importantly, the savings would "make the United States and the world safer and more secure" by creating a more efficient, less overstretched military force focused on ensuring the physical security of the US. Eliminating the further development of nuclear weapons (discussed below), space weapons, a missile defense system that does not work and is heightening tensions with Russia, and bases encircling and provoking China would further add to these savings and to our overall security.

Diplomacy, International Cooperation, and International Law. Encouragingly, many foreign policy elites have begun to embrace a new emphasis on diplomacy and international cooperation. One of the only foreign policy successes of the Bush administration—halting North Korea's nuclear program with the help of multi-nation diplomatic talks and economic concessions—is a prime example of how diplomacy can diffuse conflicts and improve national, regional, and international security. Making diplomacy the foundation, rather than an afterthought, of foreign policy begins with a substantial shift in funding from the Defense Department to the State Department and related non-military international aid programs, whose combined 2009 budget is about 30 times smaller than the Pentagon's.

The US must also recommit to full cooperation at the United Nations. While far from perfect and needing further reform, the UN is the best available forum for international cooperation, conflict resolution, and global development. Reform should include efforts to democratize the UN by expanding permanent membership in the Security Council to nations like Brazil and India, and to ensure the voice of smaller nations in decision-making processes.

A recommitment to diplomacy and international cooperation also means a recommitment to the system of international treaties, international law, and other mechanisms of global collaboration, beginning with a public pledge to uphold the Geneva conventions and prohibitions on torture (see "Human Rights" below). It should extend to joining European democracies and 106 other nations in accepting the jurisdiction of the International Criminal Court and the World Court, to joining 156 nations in the global landmine treaty, to re-signing the Anti-Ballistic Missile (ABM) treaty and forging other nuclear disarmament treaties (see "Nuclear Weapons"), and to ratifying and abiding by the Kyoto Protocol and future international agreements to confront climate change and other environmental threats (see "Climate Change").

Peacemaking and Conflict Resolution. In tandem with this emphasis on diplomacy, the US should make the peaceful resolution of international conflicts a major foreign policy goal. The successful reduction of tensions with North Korea is an important precedent and should lead to an effort to fully normalize relations between the two Koreas (which could allow the further reduction and eventual removal of all US troops from the peninsula). In the Middle East, the US should commit to local and regional peacemaking initiatives beginning with Iraq and Israel/Palestine. Just as the US should allow the UN to lead peacemaking efforts in Iraq, Foreign Policy in Focus's "Just Security" report outlines a similar plan for the Israeli-Palestinian conflict beginning with a UN-led international peace conference involving Israel, Palestinians, Arab nations, Europe, and the US.

The US should initiate another international effort aimed at decreasing equally longstanding tensions between India and Pakistan, which probably represent the most immediate threat of nuclear war on the planet. Given the US's increasingly close relations with both nations, the nation is now in a historic position, again with the assistance of the UN and other powers, to mediate Kashmir and other disagreements and to negotiate a lasting peace and demilitarization agreement. This initiative could be tied to a broader regional initiative aimed at bringing peace to Afghanistan.

Development Aid and Global Poverty Alleviation. In the face of widespread deprivation, hunger, and disease in an increasingly globalized world, the security of the US and its citizens depends on making a new commitment to dramatically increasing our development aid. While President Bush's augmentation of the foreign aid budget for HIV/AIDS treatment and other programs was again one of the few successes of his administration, Gayle Smith cites statistics showing that just 3.5 percent of 2007 national security spending went to development while 95 percent went to the military. The US currently contributes less than 0.2 percent of GNP to development assistance, the lowest among advanced industrial nations and well below the UN's target of 0.7 percent. With the help of the one-third cut in the Pentagon's budget, the US should immediately aim to increase its overseas development budget to incrementally constitute 10, then 20, then 30 percent of total national security spending, while still leaving considerable monies to ensure that domestic human security needs in education, health care, employment, and housing are also well cared for.

Importantly, this new development spending should not be directed by the Pentagon as counterinsurgency strategy and plans for AFRICOM now suggest. Instead, as many development experts are proposing, the Obama administration should create a new cabinet level foreign aid agency or significantly re-empower the US Agency for International Development (USAID) to be the main arm of US development efforts. Of equal importance, a recommitment to foreign aid demands that we ensure that assistance benefits local people and not primarily US and other western aid agencies, which now sometimes take the lion's share of aid meant to benefit others.

Ultimately, increasing development aid is not just a good and humanitarian thing to do, it will fundamentally strengthen our national security by helping to eliminate some of the economic conditions that lead to militancy and terrorism, by helping to rebuild US moral leadership and authority, and by helping our country win allies. Alternatively, says Smith, "if we fail to act now, we will be forced to pay later, both financially and with our own national security."

Humanitarian Assistance. The US must improve upon its tradition of helping others in times of humanitarian crisis. Providing emergency aid will strengthen US and global security by increasing international stability, supporting development efforts easily hindered by humanitarian disasters, and improving the global reputation of the US. This is one of the areas in which the strength and skill of the US military can best be used. As the 2004 Indian Ocean tsunami showed, the military's impressive logistics, transportation, and growing humanitarian capabilities are important tools to help people in need and to build global goodwill. These skills could be used to equal effect during domestic weather-related and geological disasters by demilitarizing elements of the US military into an unarmed emergency relief force, as part of revamping the Federal Emergency Management Agency (FEMA) in the wake of Hurricanes Katrina and Rita. Anthropologists, sociologists, and others can play an important role, particularly in efforts abroad, by ensuring that emergency relief involves the participation and guidance of locals and is provided in a socioculturally appropriate manner.

Climate Change and the Environment. Along with nuclear weapons and poverty, the threats posed by global climate change and environmental degradation are the greatest facing humanity today. For too long, the US has lagged behind other industrialized nations in reducing its environmental impact, remaining the world's biggest polluter and consumer of oil and other fossil fuels.

The US must move forcefully to develop new clean and renewable energy supplies that will ensure our long-term energy security and quickly act to reduce greenhouse gas emissions responsible for climate change. This will require major investments to develop new clean energy technologies as well as shifting tax incentives from oil production and toward encouraging energy efficiency, pollution control, and other environmental protections. Part of these efforts could come from military spending cuts (which would have the added benefit of cutting greenhouse gases that the US military produces at a rate higher than any other government institution) and through the direct conversion of armaments manufacturers into "green" industries. The US must also return to playing a leading international role in cooperatively confronting climate change and environmental degradation by signing and complying with new global agreements on emissions reductions and energy consumption.

The Abolition of Nuclear Weapons. Nuclear weapons, with their unfathomable destructive power, are the greatest threat currently facing human kind and have no place in a sane world. If we want to stop the proliferation of nuclear weapons, prevent nuclear materials from being acquired by terrorist groups, and safeguard our nation from any possible nuclear attack or accident, the United States must lead a historic effort toward the complete abolition of nuclear weapons. Some of the most prominent names in realist US foreign policy including Henry Kissinger, Sam Nunn, George Shultz, and William Perry have come to the same conclusion, arguing in a *Wall Street Journal* op-ed for "a world free of nuclear weapons." As numerous abolition plans propose, important first steps should include the cessation of all programs to upgrade US weaponry and the initiation of negotiations between the two nations with by far the largest arsenals, the US and Russia, to dramatically reduce stockpiles and take all weapons off hair-trigger alert. Subsequent negotiations involving all the nuclear powers should set goals for further reductions, rigorous monitoring and safeguards, and complete abolition, or at the very least very low numbers of weapons.

Human Rights. A human-centered foreign policy demands that the United States recommit itself to being a global leader in the protection and advancement of human rights. This must begin with correcting the scars that have so marred the nation's human rights record: the full closure of all detention facilities at Guantánamo Bay and all secret detention facilities globally; the complete repudiation of torture by the US government and all its employees and contractors, including the CIA; the termination and condemnation of extraordinary rendition as a practice sending detainees to countries known to use torture as an interrogation technique; and a reaffirmation of the US's full commitment to the Geneva conventions and all other applicable international conventions governing torture, abusive treatment, and conduct during war.

Next the US should return to its place as a leader in promoting the extension and protection of human rights globally, including social, economic, and political rights. Again this starts at home by working to bring our domestic human rights record into full compliance with all applicable international human rights covenants, beginning with the rights of women, ethnic and other minorities, immigrants, and prisoners.

As the nation corrects its own human rights record, we should begin to encourage human rights compliance in other nations, including the expansion of democratic rights, as part of our broader diplomatic relations.

A Guide for Action

While we must reject counterinsurgency and the entire Bush neoconservative legacy as the tragic strategies of empires past, these proposals for a humanpolitik are not a call for isolationism or disengagement. In our globalizing world, our security is directly connected to the security of others and demands mutual engagement and cooperation on the world's problems. Despite the destructive impact of Bush's foreign policy, for at least a decade we have seen the stirrings of a broadening public involvement in US foreign relations and in the global relations of peoples and nations worldwide—most notably in global anti-war protests that showed far more wisdom about a war in Iraq than most supposed foreign policy experts in the US. To achieve the fundamental shift in US foreign policy that needs now to occur, citizens must play ever more active roles in global relations by envisioning and demanding a new US foreign policy, researching policy alternatives, resisting warmaking, building transnational cooperation at every level, and by helping to make the world a more just, equitable, and secure place for all human beings. Anthropologists, in particular, often so good at critique, must work harder to research, develop, and implement foreign policy alternatives, to inform and influence policymakers, and to use their cross-cultural communication skills to assist in efforts for peaceful international cooperation.

Four decades ago, as the nation was deepening its involvement in Vietnam, a former National Security Council staffer wrote to *The New York Times* urging an end to "a needless war." "For the greatest power on earth has the power denied to others," James C. Thompson, Jr. said.

> The power to take unilateral steps, and to keep taking them; the power to be as ingenious and relentless in the pursuit of peace as we are in the infliction of pain; the power to lose face; the power to admit error, and the power to act with magnanimity.

We still have that power, though we may not for long. What we choose to do with the power hangs in the balance.

Pledge of Non-participation in Counterinsurgency

We, the undersigned, believe that anthropologists should not engage in research and other activities that contribute to counterinsurgency operations in Iraq or in related theaters in the "war on terror." Furthermore, we believe that anthropologists should refrain from directly assisting the US military in combat, be it through torture, interrogation, or tactical advice.

US military and intelligence agencies and military contractors have identified "cultural knowledge," "ethnographic intelligence," and "human terrain mapping" as essential to US-led military intervention in Iraq and other parts of the Middle East. Consequently, these agencies have mounted a drive to recruit professional anthropologists as employees and consultants. While often presented by its proponents as work that builds a more secure world, protects US soldiers on the battlefield, or promotes cross-cultural understanding, at base it contributes instead to a brutal war of occupation which has entailed massive casualties. By so doing, such work breaches relations of openness and trust with the people anthropologists work with around the world and, directly or indirectly, enables the occupation of one country by another. In addition, much of this work is covert. Anthropological support for such an enterprise is at odds with the humane ideals of our discipline as well as professional standards.

We are not all necessarily opposed to other forms of anthropological consulting for the state, or for the military, especially when such cooperation contributes to generally accepted humanitarian objectives. A variety of views exist among us, and the ethical issues are complex. Some feel that anthropologists can effectively brief diplomats or work with peacekeeping forces without compromising professional values. However, work that is covert, work that breaches relations of openness and trust with studied populations, and work that enables the occupation of one country by another violates professional standards.

Consequently, we pledge not to undertake research or other activities in support of counterinsurgency work in Iraq or in related theaters in the "war on terror," and we appeal to colleagues everywhere to make the same commitment.

_____

_____

_____

Signed pledges can be submitted to:
Network of Concerned Anthropologists
Sociology and Anthropology Department
George Mason University, MSN-3G5
Robinson Hall B305
4400 University Drive
Fairfax, VA 22030

About the Contributors

Catherine Besteman is Professor of Anthropology at Colby College. Her work examines post-apartheid transformations in South Africa, with a particular focus on local activists working to overcome Cape Town's enduring patterns of racism and poverty. Her latest book is *Transforming Cape Town* (California, 2008).

Andrew Bickford is Assistant Professor of Anthropology and Sociology at George Mason University. He is writing a book on health and technology in the US military, entitled *Disposable, Deployable, Forgettable: The Paradoxes of Health in the United States Military*.

Greg Feldman is Assistant Professor of International Migration at the University of British Columbia. His book manuscript is entitled *Managing Migrants: Security and Labor in an Age of European Demographic Decline* (Stanford, under contract).

Roberto González is Associate Professor of Anthropology at San Jose State University. He is the author of *Zapotec Science: Farming and Food in the Northern Sierra of Oaxaca* (Texas, 2001) and *American Counterinsurgency: Human Science and the Human Terrain* (Prickly Paradigm, 2008).

Hugh Gusterson is Professor of Anthropology and Sociology at George Mason University. He is the author of *Nuclear Rites* (California, 1996) and *People of the Bomb* (Minnesota, 2004) and writes a regular online column for the Bulletin of Atomic Scientists. With Catherine Besteman he co-edited *Why America's Top Pundits Are Wrong* (California, 2005) and *The Insecure American* (California, 2009).

Kanhong Lin is a graduate student of anthropology at San Jose State University. He is the co-author of the 2006 American Anthropological Association's resolution condemning torture and its use by US forces.

Catherine Lutz is a Watson Institute Professor at Brown University, where she holds a joint appointment with the Department of Anthropology. Her most recent books include *Local Democracy under Siege: Activism, Public Interests, and Private Politics* (New York University, 2007) and *Homefront: A Military City and the American 20th Century* (Beacon, 2001), which won the Leeds Prize and the Victor Turner Prize.

David Price is Professor of Anthropology and Sociology at St. Martin's University. He is the author of *Threatening Anthropology: McCarthyism and the FBI's Surveillance of Activist Anthropologists* (Duke, 2004) and *Anthropological Intelligence: the Deployment and Neglect of America Anthropology in the Second World War* (Duke, 2008).

David Vine is Assistant Professor of Anthropology at American University. He is the author of the book *Island of Shame: The Secret History of the U.S. Military Base on Diego Garcia* (Princeton, 2009). His other writing has appeared in *The New York Times*, *The Washington Post*, *Mother Jones* online, *Foreign Policy in Focus*, *The Chronicle of Higher Education*, and *International Migration*, among others.

Also available from Prickly Paradigm Press:

Paradigm 1   *Waiting for Foucault, Still*
             Marshall Sahlins

Paradigm 2   *War of the Worlds: What about Peace?*
             Bruno Latour

Paradigm 3   *Against Bosses, Against Oligarchies: A Conversation with Richard Rorty*
             Richard Rorty, Derek Nystrom, and Kent Puckett

Paradigm 4   *The Secret Sins of Economics*
             Deirdre McCloskey

Paradigm 5   *New Consensus for Old: Cultural Studies from Left to Right*
             Thomas Frank

Paradigm 6   *Talking Politics: The Substance of Style from Abe to "W"*
             Michael Silverstein

Paradigm 7   *Revolt of the Masscult*
             Chris Lehmann

Paradigm 8   *The Companion Species Manifesto: Dogs, People, and Significant Otherness*
             Donna Haraway

Paradigm 9   *9/12: New York After*
             Eliot Weinberger

Paradigm 10 *On the Edges of Anthropology (Interviews)*
             James Clifford

Paradigm 11 *The Thanksgiving Turkey Pardon, the Death of Teddy's Bear, and the Sovereign Exception of Guantánamo*
             Magnus Fiskesjö

Paradigm 12 *The Root of Roots: Or, How Afro-American Anthropology Got its Start*
             Richard Price and Sally Price

Paradigm 13 *What Happened to Art Criticism?*
             James Elkins

Paradigm 14 *Fragments of an Anarchist Anthropology*
             David Graeber

continued

Paradigm 15 *Enemies of Promise: Publishing, Perishing, and the Eclipse of Scholarship*
Lindsay Waters

Paradigm 16 *The Empire's New Clothes: Paradigm Lost, and Regained*
Harry Harootunian

Paradigm 17 *Intellectual Complicity: The State and Its Destructions*
Bruce Kapferer

Paradigm 18 *The Hitman's Dilemma: Or, Business, Personal and Impersonal*
Keith Hart

Paradigm 19 *The Law in Shambles*
Thomas Geoghegan

Paradigm 20 *The Stock Ticker and the Superjumbo: How The Democrats Can Once Again Become America's Dominant Political Party*
Rick Perlstein

Paradigm 21 *Museum, Inc.: Inside the Global Art World*
Paul Werner

Paradigm 22 *Neo-Liberal Genetics: The Myths and Moral Tales of Evolutionary Psychology*
Susan McKinnon

Paradigm 23 *Phantom Calls: Race and the Globalization of the NBA*
Grant Farred

Paradigm 24 *The Turn of the Native*
Eduardo Viveiros de Castro, Flávio Gordon, and Francisco Araújo

Paradigm 25 *The American Game: Capitalism, Decolonization, World Domination, and Baseball*
John D. Kelly

Paradigm 26 *"Culture" and Culture: Traditional Knowledge and Intellectual Rights*
Manuela Carneiro da Cunha

Paradigm 27 *Reading* Legitimation Crisis *in Tehran: Iran and the Future of Liberalism*
Danny Postel

Paradigm 28 *Anti-Semitism and Islamophobia: Hatreds Old and New in Europe*
Matti Bunzl

Paradigm 29 *Neomedievalism, Neoconservatism, and the War on Terror*
Bruce Holsinger

Paradigm 30 *Understanding Media: A Popular Philosophy*
Dominic Boyer

Paradigm 31 *Pasta and Pizza*
Franco La Cecla

Paradigm 32 *The Western Illusion of Human Nature: With Reflections on the Long History of Hierarchy, Equality, and the Sublimation of Anarchy in the West, and Comparative Notes on Other Conceptions of the Human Condition*
Marshall Sahlins

Paradigm 33 *Time and Human Language Now*
Jonathan Boyarin and Martin Land

Paradigm 34 *American Counterinsurgency: Human Science and the Human Terrain*
Roberto J. González

Paradigm 35 *The Counter-Counterinsurgency Manual: Or, Notes on Demilitarizing American Society*
Network of Concerned Anthropologists

Paradigm 36 *Are the Humanities Inconsequent? Interpreting Marx's Riddle of the Dog*
Jerome McGann